Russian Oil Enterprises in Europe

Tomáš Vlček • Martin Jirušek

Russian Oil Enterprises in Europe

Investments and Regional Influence

Tomáš Vlček
Faculty of Social Studies
Masaryk University
Brno, Czech Republic

Martin Jirušek
Faculty of Social Studies
Masaryk University
Brno, Czech Republic

ISBN 978-3-030-19838-1   ISBN 978-3-030-19839-8   (eBook)
https://doi.org/10.1007/978-3-030-19839-8

This Palgrave Macmillan imprint is published by the registered company Springer Nature Switzerland AG
The registered company address is: Gewerbestrasse 11, 6330 Cham, Switzerland

ANNOTATION

In this book, the authors address popular accusations that Russia misuses energy supplies to help achieve its foreign policy goals. To determine whether these accusations are well grounded, the authors constructed a specific research model that was then applied to the states of South-Eastern Europe, an area believed to be a principal Russian focus. The book concentrates on the oil sector, which, due to its complexity, may be seen as a breeding ground for politically influenced deals and backroom negotiations. The book offers 11 thorough case studies that examine the oil sectors of individual countries in South-Eastern Europe with regard to the activities of Russian companies and representatives. As a result, the book unearths the key patterns of conduct by Russian companies and the role played by the Russian state. In doing so, it answers the question of whether the Russian state employs energy supplies as a foreign policy tool. To provide a comprehensive picture of the environment and to unveil potentially more complex behavioural patterns, the book is supplemented with a chapter focusing on the natural gas sector. It thus provides a comprehensive overview of the environment that is useful for scholars as well as analysts and decision-makers.

Section 3.7 in Chap. 3, and Chaps. 4, 5, 6, 7, 8, 9, 10, 11, 12, and 13 are written by Tomáš Vlček; Chaps. 3 (excl. Sect. 3.7), 14, and 15 are written by Martin Jirušek; and Chaps. 1 and 2 are written by Tomáš Vlček and Martin Jirušek.

ACKNOWLEDGEMENTS

The authors would very much like to thank the numerous experts, partners, and colleagues who were willing to meet with us and consult the energy security and energy policy of the countries of South-Eastern Europe. Even when they were either unwilling or unable to go on record, they openly discussed the topics with us and considerably aided our understanding of both context and particularities.

The authors are also thankful for support from the Masaryk University Specific Research Programme titled Europe in the Changing International Environment IV (MUNI/A/0834/2017). Similarly, the authors wish to thank their colleagues both inside and outside of Masaryk University who helped by providing commentaries or insights. This book could not have been written without the support provided by the university and individuals in various faculty bodies. Namely, the authors would like to thank Břetislav Dančák, Vít Hloušek, and Petr Suchý.

Finally, the authors would like to express their gratitude to Mark Alexander, who proofread the monograph.

Contents

LIST OF FIGURES

LIST OF TABLES

CHAPTER 1

Introduction

Over the last decade, relations between Russia and the West have sustained substantial damage, and energy-related disputes have contributed to this state of affairs to a considerable degree. These disputes have caused memories to resurface of an era when energy was used as a foreign policy tool, one that catalysed serious worldwide economic slumps. In past years, Russia has been accused of misusing its prominence as an energy supplier for political leverage, putting pressure on countries that depend upon Russia for their energy supplies.

The widespread perception is that Russia sees energy exports as a legitimate foreign policy tool, and that the strategy of leading Russian energy companies is driven by government directives that may be geopolitically rather than economically motivated. This perception often dominates the debate, especially in the countries of Central, Eastern, and South-Eastern Europe (SEE). Russia's handling of energy resources is particularly sensitive for states that depend on Russian supplies, infrastructure, or technologies. An added factor is Russia's strong opposition to the expansion of Western economic, political, and security organizations that came with the dissolution of the Eastern Bloc at the end of the 1980s. Its general policy aim was to preserve the neutrality of post-communist states, so they could serve as a kind of 'bumper region' between East and West (as was suggested for Central Europe [CE]), but this did not come to pass; membership in Western organizations was more attractive to the former communist states. The way the ensuing transitional process played out was far from

© The Author(s) 2019
T. Vlček, M. Jirušek, *Russian Oil Enterprises in Europe,*
https://doi.org/10.1007/978-3-030-19839-8_1

even for all concerned countries, though. The transition path for states in the SEE was far less straightforward than it was for the states of Central Europe (CE). Simultaneously, Russia took a stronger hand in SEE than it did in CE, partly because of geographical proximity and, in some cases, due to political, historical, or cultural ties. The bottom line is that it was generally accepted that Russia might be willing to use its energy resources coercively to steer development within individual countries or in the region at large.

Because of their importance for economies, the spotlight has understandably been on hydrocarbons, but most of the attention has fallen on natural gas, which made headlines in January 2006 and 2009, and was closely interlinked with the crisis in relations between Russia and Ukraine. Chiefly due to market liquidity factors, oil has been overlooked in discussions of political pressure. It is nevertheless an area worth looking at. The comparatively more fragmented, less predictable nature of the sector versus natural gas may be a breeding ground for politically influenced deals and backroom negotiations at a more granular level which, however, may permit a more complex, less transparent network through which to exert pressure.

Although natural resources have always accounted for a significant percentage of Russia's exports, they came to play an even bigger role after the year 2000. In the preceding decade, oil exports rose significantly. Along with rising oil prices, the exports were responsible for a much-needed influx of capital into the Russian state budget. Natural gas exports also rose sharply, driven by greater demand and those rising oil prices. During this era, oil and gas exports became Russia's most important export commodity, and they have remained so ever since. From this perspective, it is no wonder the Russian government pays great attention to developments in the energy markets, since they know that any changes which occur there may substantially impact the Russian economy. With the principal energy companies in the hands of the state, the government theoretically has the tools to influence state energy policy and to use these companies to achieve its aims, be they political or economic. At the same time, the companies find themselves in a somewhat untenable position in that any decline in energy sales could have major consequences not only for their own bottom line but also for the producer's economy. They must therefore tread carefully so as not to aggravate their customers. Failure to do so could lead to accusations that energy is being misused to further state goals, seriously impacting future sales and harming the economy.

In discussing concerns over Russia's potential misuse of its position as a key energy supplier, it is imperative to note the changes that have taken place in the relationship between Russia and Europe over the past several years. Since Vladimir Putin assumed the presidency, the Russian government has implemented measures to increase government involvement in various elements of the administrative apparatus, including those involved with the energy sector. Over the past 15 years, several steps have been taken to help restore state control over the most important parts of the sector. Within the past decade, several disputes have also arisen in Europe in which Russian energy sources played a vital role. But the cumulative impact of all these events has been to arouse concerns in those states that depend on Russia for their energy supplies.

South-Eastern Europe is an interesting region with a colourful yet often troubled past. A defining characteristic of the region has been its delayed economic transformation, which stems largely from its geographical location and its political and historical development. The region also has a track record of unique relationships with the Soviet Union and later Russia which, in turn, perceives these countries to be unique. As such, the energy sector too has taken an uncommon path. A number of the countries are closely bound to Russia in terms of supply, infrastructure, and technology and are thereby susceptible to pressure. This susceptibility is further highlighted by the pronounced role played by the industrial sector in the countries' overall economic output. Their economic structure was considerably influenced by the Soviet model and its focus on industrial production and postwar reconstruction. Because of this, the economies of the former communist states in SEE are highly energy intensive and, consequently, supply sensitive. As a result, any politicization of energy supplies naturally raises concerns. Importantly, such politicization has already taken place a number of times. Many states within the region have experienced the coupling of energy-related issues to politics, and Russia has been accused on several occasions of using energy supplies as a foreign policy tool.

In this book, the authors address the question of whether Russia really misuses oil supplies for its policy goals and what conditions surround such use. To address the research aim, the following research question was formulated: 'Do Russian state-owned energy companies in the oil sector behave like tools of the Russian state in South-Eastern European countries, and do they serve as vehicles of Russian foreign policy?' The research is based in the realist paradigm in international relations which gave birth to the so-called strategic approach to energy policy, emphasizing geopolitical logic and the importance of energy resources for state power and their use as foreign policy tools. For purposes of this research, the authors developed an ideal type model of state-guided, strategically oriented behaviour that is characterized by a set of features and indicators. These directly target the main characteristics displayed in attempts to politicize energy supplies, precisely what Russia has been accused of doing in various cases. A search was then made for these indicators in individual cases/ states to assess the extent to which Russian state-owned oil companies and their home state engage in behaviour perceived to be problematic.

The focus of the book is on the South-Eastern European region, comprising 11 states: Romania, Bulgaria, Greece, Albania, Macedonia, Kosovo, Serbia, Montenegro, Bosnia and Herzegovina, Croatia, and Slovenia. The majority of these states share a similar historical experience of authoritarian regimes, more or less bound to the former Soviet Union, which influenced the internal structure of their economies, including the energy sector. As much as the region might be perceived as a more or less homogeneous group of states experiencing a delayed economic transition (with the exception of Greece, Slovenia, and probably also Croatia), the reality on the more granular level is somewhat different. Some states have entered the EU, some are candidate states. Some still struggle with basic economic reforms, while others have emerged from the transitional period in good shape. This diversity itself provides a great incentive for undertaking research. The region's importance from a European energy security standpoint, its interesting internal dynamics, and the high level of Russian involvement on one side, combined with the unclear relationship between Russian foreign policy and the conduct of Russian energy companies on the other, are the main motivators for the research and the book respectively.

Russia's perception of the region is also worth paying attention to, not just because of the currently worsening state of relations between Russia and the West, but also from the perspective of Russia's long-term stance towards the region. In contrast to the CE region, where Russia abandoned its former positions during the initial stages of the region's reorientation towards the West, in SEE and to an even greater extent in some of the Balkan states, Russian foreign policy has been somewhat touchier. The peculiar nature of the relationship has been demonstrated on more than one occasion, including Russia's stance during the Balkan wars and its attitude towards the recognition of Kosovo. Russia's economic involvement has also been more intensive, and the energy sector is no exception. Accusations of nonstandard deals, coupled with cultural proximity and close ties between some Russian and local politicians, definitely beg for thorough examination.

The book also includes a brief analysis of the situation in the natural gas sector with a similar goal in sight. The main intention is to compare and contrast the situation in both sectors to determine the extent to which the behavioural patterns are similar. The findings thus help to understand the general code of conduct employed by Russian energy companies and their potential ties to their home government.

The uniqueness of the book stems from the evidence-based approach it takes. It is built on field research conducted in the countries in question and on a unique theoretical model of behaviour. As such, not only does it come up with highly specific findings, it also provides a basis upon which scholars and decision-makers may assess individual cases in the years to come. The set of in-depth analyses, which provides a thorough understanding of individual countries, possesses value on its own, especially for the Western Balkans, whose states have been largely neglected in energy-related research so far.

The book has been structured into 15 chapters, starting with this Introduction. In Chap. 2, the authors review the current literature, discovering that even though Russia's foreign policy and its (foreign) energy policy are important and frequent topics in current research, its relationship to South-Eastern Europe has been largely neglected.

In Chap. 3, the authors present their research design, starting with the theoretical framework and progressing to the creation of the model employed throughout the book. The chapter thus includes the theoretical basis for the model and the model itself. The chapter also describes which Russian oil companies are under scrutiny and why PAO Lukoil, even though privately owned, was included in the research.

Chapters 4, 5, 6, 7, 8, 9, 10, 11, and 12 contain a set of 11 detailed case studies of Albania, Bosnia and Herzegovina, Bulgaria, Croatia, Greece, Macedonia, Romania, Serbia, and countries with limited Russian activities—Kosovo, Montenegro, and Slovenia. These chapters follow an identical structure, starting in each case with an introduction and then presenting the upstream, midstream, and downstream sections of the country's oil value chain. These detailed sections provide much-needed insight into the oil sector of the countries and are followed with information on Russian involvement and operations in their oil markets. Each chapter then concludes with an evaluation of the research indicators showing Russian activity in the oil sector of the country.

Chapter 13 summarizes the findings of Chaps. 4, 5, 6, 7, 8, 9, 10, 11, and 12 and presents the overall outcomes for the oil sector in a succinct but coherent manner. Russian interests, policies, and activities are explained in their historical, economic, and strategic contexts.

Chapter 14 presents the findings of research into the conduct of the company Gazprom. The findings are presented in a condensed fashion, sketching out the behavioural patterns of the Russian gas giant in the South-Eastern European region. By doing this, building on the same methodology, the chapter complements the chapters focused on the oil sector.

A short, highly condensed general conclusion is presented in Chap. 15. The findings are put in a broader context of the South-Eastern Europe-Russia relations.

The authors are confident in saying that the book as a whole provides a comprehensive picture of Russian oil and gas companies and their individual strategies in an important European region and thus contributes to the broader geopolitical discussion. We believe that such an evidence-based discussion is what the field needs, especially in our contemporary world of turbulent, less predictable interstate relations.

At the conclusion of this introductory chapter, the authors would like to state that some chapters build upon their previously published research or expand their long-term research interests. With the permission of the respective publishing houses, we have used small sections from the previously published books *Politicization in the Natural Gas Sector in South-Eastern Europe: Thing of the Past or Vivid Present?* (Jirušek, 2017), *Challenges and Opportunities of Natural Gas Market Integration in the Danube Region: The South-West and South-East of the Region as Focal Points for Future Development* (Jirušek & Vlček, 2017), *Energy Security in Central and Eastern Europe and the Operations of Russian State-Owned Energy Enterprises* (Jirušek et al., 2015), and *Alternative Oil Supply Infrastructures for the Czech Republic and Slovak Republic* (Vlček, 2015). These books, published by Masaryk University Press, were consulted and made partial use of in Chaps. 3, 7, 13, and 14. A paper entitled 'Russia's Energy Relations in Southeastern Europe: An Analysis of Motives in Bulgaria and Greece' (Jirušek, Vlček, & Henderson, 2017) and published by Taylor & Francis in the journal *Post-Soviet Affairs* was consulted and partially used in Chaps. 3, 6, 8, and 14. Finally, a paper entitled 'The Hydrocarbons Sector in Albania: Short-Term Challenges and Long-Term Opportunities' (Vlček & Jirušek, 2018), published by Duke University Press in the journal *Mediterranean Quarterly*, was consulted and used in Chap. 4.

Sources

Jirušek, M. (2017). *Politicization in the Natural Gas Sector in South-Eastern Europe: Thing of the Past or Vivid Present?* Brno: Masaryk University. ISBN 978-80-210-8881-8.

Jirušek, M., & Vlček, T. (2017). *Challenges and Opportunities of Natural Gas Market Integration in the Danube Region. The South-West and South-East of the Region as Focal Points for Future Development.* Brno: Masaryk University. ISBN 978-80-210-8750-7. Retrieved from https://munispace.muni.cz/library/catalog/book/945

Jirušek, M., Vlček, T., & Henderson, J. (2017). Russia's Energy Relations in Southeastern Europe: An Analysis of Motives in Bulgaria and Greece. *Post-Soviet Affairs, 33*(5), 335–355. https://doi.org/10.1080/1060586X.2017.1341256. Retrieved from www.tandfonline.com

Jirušek, M., Vlček, T., Koďousková, H., Robinson, R. W., Leshchenko, A., Černoch, F., et al. (2015). *Energy Security in Central and Eastern Europe and the Operations of Russian State-Owned Energy Enterprises.* Brno: Masaryk University. ISBN 978-80-210-8048-5. Retrieved from https://munispace.muni.cz/library/catalog/book/790

Vlček, T. (2015). *Alternative Oil Supply Infrastructures for the Czech Republic and Slovak Republic.* Brno: Masaryk University. ISBN 978-80-210-8035-5.

Vlček, T., & Jirušek, M. (2018). The Hydrocarbons Sector in Albania: Short-Term Challenges and Long- Term Opportunities. *Mediterranean Quarterly, 29*(1), 96–119. https://doi.org/10.1215/10474552-4397358

CHAPTER 2

Literature Review

Energy policy has to a large extent been overlooked in the existing literature except in work dealing with the history of international relations and geopolitics, where it, along with energy supply, is often treated as a subcategory of foreign policy. By nature this involves simplifying the issue to simple 'weaponization' of energy supplies. This understanding is not outright wrong, but we maintain there is more to the issue, particularly given that in recent years energy security is increasingly being perceived as an independent area of research.

Among the most commonly cited works, those that have laid the foundations of the discipline, are Michael S. Hamilton's *Energy Policy Analysis: A Conceptual Framework* (Hamilton, 2012) and Gawdat Bahgat's *Energy Security: An Interdisciplinary Approach* (Bahgat, 2011). The latter enumerates the most important issues and actors while elaborating on the main sticking points between producers and consumers. A similarly comprehensive treatment, albeit one that focuses on contemporary issues, is offered in *Energy and Security: Strategies for a World in Transition* by Jan H. Kalicki and David L. Goldwyn (2013). *Energy Security and Global Politics: The Militarization of Resource Management*, edited by Daniel Moran and James A. Russell (Moran & Russell, 2008), is also a foundational work and is typical in highlighting the importance of oil and gas, a cornerstone of most of the literature on energy security. Further reading includes *Energy Security: Europe's New Foreign Policy Challenge* by Richard Youngs (2009) and Luft and Korin (2009). Although we maintain

© The Author(s) 2019
T. Vlček, M. Jirušek, *Russian Oil Enterprises in Europe*,
https://doi.org/10.1007/978-3-030-19839-8_2

that books with a historical focus lack the analytical rigour to penetrate all layers of the issue (see further), we do appreciate them as necessary sources of underlying information vital for the analysis at large. A comprehensive take on the history of Russian gas exports to Europe by Per Högselius (2013) or the work of Daniel Yergin (2008, 2012) on the history of oil industry is worth mentioning in this regard.

A plethora of literature exists about the secondary focal point of this book, Russia,[1] much of it written from diverse perspectives. Here we may start with broader work that focuses on the Russian role in international relations and the tools Russia employs in the foreign policy sphere. Most books in this area have a history focus, concentrating on high-level politics in which Russia has been playing an integral role in shaping the world and the European order. This group includes *Russia's Foreign Policy: Change and Continuity in National Identity* by Andrei P. Tsygankov (2010), and *Russian Foreign Policy: The Return of Great Power Politics* by Jeffrey Mankoff (2009). While the first book focuses on the main determining factors of Russian foreign policy and their potential ramifications in real-life politics, the latter takes on recent Russian activities in the field of international relations, focusing on the era of Vladimir Putin.[2] Mankoff's book is also useful in interconnecting the spheres of foreign policy and energy policy. He asserts that Russia is in the process of reassuming its position as a key international player and partly levers its vast energy resources to do so—a common yet rarely verified claim in the contemporary literature. Geopolitical reasoning and a geopolitical explanation for Russia's foreign policy are also very frequent, as evidenced in some outstanding works, some quite recent (Kaplan, 2012, pp. 154–187; Kirchick, 2017, pp. 11–39; Marshall, 2015, pp. 11–39). Other related publications that deserve mention include Øistein Harsem and Dag Harald Claes' discussion of the interdependence of EU and Russian energy relations (Harsem & Claes, 2013); analysis of the energy relations and diplomacy of EU and Russia (Aalto, 2007); David Svarin's analysis of the construction of 'geopolitical spaces' in Russian foreign policy discourse (Svarin, 2016); Martin Jirušek's and Petra Kuchyňková's research revealing Gazprom has behaved in such a way as to indicate that it is being used as a tool of foreign policy, but the

---

[1] Here, the term 'Russia' refers to state units existing within or overlapping with the area of the current Russian Federation.

[2] By the term 'era', the authors refer to the period which started when Vladimir Putin first assumed the presidency, through his years as prime minister, to the present.

primary factor controlling its behaviour remains the environment in which the company is operating (Jirušek & Kuchyňková, 2018); and Andrei P. Tsygankov's complex work on Russia's foreign policy (Tsygankov, 2016). Vladimir Putin himself is an object of interest in many scholarly books (Dawisha, 2015; Hill & Gaddy, 2015; Myers, 2015; Salminen, 2015; Taylor, 2018, and many others).

However, as hinted above, the authors are of the opinion that this notion has gotten far too much attention in the literature and oversimplifies the way Russia treats energy supplies. Scholars tend to perceive energy supplies of Russian origin as straight-off tools or outright weapons conventionally used in Russian foreign policy. Energy, especially oil and natural gas, is used in their view for foreign policy leverage (Considine & Kerr, 2002; Hill, 2004; Morse & Richard, 2002; Newnham, 2011; Stern, 2005, etc.). An interesting contribution to the topic is a paper by Harley Balzer, who connects policies enacted during Vladimir Putin's second presidential term with the views expressed in Putin's dissertation and a published article (Balzer, 2005).

Although we are far from rejecting this realist explanation as outright nonsense, we must maintain scientific rigour and derive our outcomes from the evidence and verifiable findings. It seems that authors especially tend to treat energy exports[3] as an indispensable part of the state's toolbox. Robert W. Orttung and Indra Overland (2010), for example, argue that Russia pursues 'a rational set of political and economic goals in its foreign energy policy, but that it is constrained in its efforts by the set of tools available to it'. Mert Bilgin is of the opinion that control over markets and supplies is directly related to the use of natural gas as leverage in Russian foreign (energy) policy (Bilgin, 2011). The relative simplification may be caused by the use of a selective approach to the issue and a tendency to elevate anecdotal evidence and extreme examples to the level of universal explanations.

There is thus a multitude of authors not in agreement with the realist explanation of Russian foreign (energy) policy. Elena Kropatcheva (2012) argues that Russian foreign policy is selective and includes both cooperative and non-cooperative tactics; thus, accusations that Russia is anti-Western and unwilling to cooperate are unfounded. Shortcomings of the contemporary debate over Russia's energy weapon made evident by the mixed diplo-

---

[3] The term 'energy exports' may also include other energy-related commodities like technologies or know-how.

macy and outcomes in these successive gas crises are discussed by Adam N. Stulberg (2015). Andreas Goldthau is of the opinion that the threat to the European gas supply lies not in geopolitics, but rather in a lack of investment in the Russian upstream sector (Goldthau, 2008). Pami Aalto and his colleagues analysed Russia's energy policies from different levels of analysis: national, interregional, and global (Aalto, 2012). Strict limits on the energy weapon, whoever employs it, were noted by Andreas Goldthau and Tim Boersma (2014). Andrew Monaghan in his paper suggested that Russian energy diplomacy is flawed by the lack of a coherent, consensus-based strategy for achieving the superpower status. He further suggested that in Russia's case, weakness goes together with strength and obliges Russia to be cooperative in its foreign relationships (Monaghan, 2007). The list of authors who disagree with the use of the term 'energy weapon' or with the realist explanation of Russian foreign (energy) policy continues: Tatiana Romanova (2014), Thijs Van de Graaf and Jeff D. Colgan (2017), Roland Dannreuther (2016), and others. Probably the most detailed analysis of the term comes from Karen Smith Stegen (2011). In a widely read article, she maintains that in the majority of cases, Russian manipulation failed to achieve political concessions. Only in some cases does Russia seem to have succeeded in making successful use of its energy weapon (the Black Sea). Stegen further argues that 'the evidence for the consistently successful implementation of the energy weapon by Russia is less than overwhelming' as 'client states, even weak and highly dependent states such as the Baltic countries and Georgia, were able to resist changing their policies to appease Russia, often through the use of strategic alliances' (Stegen, 2011, p. 6511).

Since this book is case- and region-specific, our aim is to provide a thorough examination of a spatially limited population of cases based on evidence and field research. We therefore refrain from making premature statements or judgements based on a handful of examples that are not entirely representative. Rather, we derive our outcomes from the entire population of cases within the area examined.

Now to the relevant energy-related literature. Understanding the global energy environment, the fundamentals of the oil and gas industry, and particularly the role played by Russia are essential. Therefore, the reader should focus on books like *Oil 101* by Morgan Downey (2009). Additional reads include *Oil & Gas Company Analysis: Upstream, Midstream and Downstream* (Colombano & Colombano, 2015) and *The Global Oil & Gas Industry: Management, Strategy & Finance* by Andrew Inkpen and Michael H. Moffett (2011) that also focus on natural gas, which has been traditionally perceived as a commodity intertwined with oil. But although these works do offer a basic explanation of the inner processes within these sectors that is crucial for analysis, they barely touch on the issue of the politicization of supplies. In this sense, *Oil Politics* by Francisco Parra (2010) serves better. Still, the overview provided of key events related to the historical development of the global oil environment is fairly dia-chronic in nature.

In speaking of Russia's role in supplying energy to the region and to Europe as a whole, given the current situation in the international system it is also important to consider the ties between Russia's energy sector and its government. It is nothing new to state that there are close ties between the Russian state and the Russian energy sector; this is determined by the fact that natural resources (oil and gas in particular) form a crucial source of funding for the Russian state budget. A theoretical grounding for assessing a state's behaviour with a view to its economic goals may be found in a comprehensive book by Robert Gilpin, *Global Political Economy: Understanding the International Economic Order* (Gilpin, 2001). Publications focusing specifically on the Russian oil and gas industry and role in the state's administration mainly include reports like *Key Determinants for the Future of Russian Oil Production and Exports* (Henderson, 2015a) published by the Oxford Institute for Energy Studies, a prominent institute focusing on the issue of energy policy and related matters. The institute is known for its large portfolio of works on energy policy and security, includ-ing, among others, *The Russian Gas Matrix: How Markets Are Driving Change*, also by James Henderson and Simon Pirani (Henderson & Pirani, 2014). Other publications include, for example, papers by Christian Dreger, Konstantin A. Kholodilin, and Dirk Ulbricht and Jarko Fidrmuc (2016) analysing the importance of oil to Russia and indicating that the bulk of the ruble's recent depreciation may be attributed to the decline in oil prices as opposed to the impact of the economic sanctions imposed after the Crimea annexation. Adnan Vatansever examines Russia's entire oil and gas export network and reveals that there is considerable surplus pipeline capacity, offering enough manoeuvring room that, in the oil sector, Russia would be free to abandon an entire route of its choice (Vatansever, 2017). The sig-nificance of the historically problematic downstream sector for the Russian petroleum industry is scrutinized by Nikita O. Kapustin and Dmitry A. Grushevenko (2018), who offer an indirect answer to the question of why refineries located in foreign countries are such an important asset for Russian companies.

Taken together, the publications in the previous paragraphs sketch out a convincing picture of the foundational importance of oil and gas for the Russian economy. They include hints at using energy supplies as tools, as explained by Gilpin (see above), but there are other publications as well that address the topic and are worth mentioning. Given the role played by Putin in the Russian state and the state's role in the natural gas and oil sectors—which has been even more prominent since the moment the current president took office—it is hardly any wonder that monographs have focused on Putin's part in Russia's foreign and energy policy. One such book is titled *The New Cold War: Putin's Russia and the Threat to the West*, by the renowned journalist Edward Lucas (2014). Further, Bertil Nygren, in his book *The Rebuilding of Greater Russia: Putin's Foreign Policy Towards the CIS Countries* (Nygren, 2007), analyses Russia's foreign policy towards the former countries of the Soviet Union and clearly indicates that energy policy and supplies of vital energy commodities do play an important role. The book may also serve as a transition to a similarly focused works dealing with Russia and exerting power but concentrating more on actual energy-related ties between Russia and other states or regions. This group includes works by Margarita M. Balmaceda (2012, 2015), David G. Victor, Amy M. Jaffe, and Mark H. Hayes (2008), Jeronim Perović, Robert W. Orttung, and Andreas Wenger (2009), Adrian Dellecker and Thomas Gomart (2011), Andrey A. Konoplyanik (2012), Philipp M. Richter and Franziska Holz (2015), Matúš Mišík and Veronika Prachárová (2016), or András Deák (2014). Additional readings include papers by Steven Woehrel (2012), Anke Schmidt-Felzmann (2011), Jan Osička and Petr Ocelík (2017), Ivan Tchalakov and Tihomir Mitev (2019), James Henderson (2015b), or Alexander Gusev and Kirsten Westphal (2015).

An area-focused book that examines energy-related issues in the region is *Rival Power: Russia in Southeast Europe*, by Dimitar Bechev (2017). Although the book mentions natural gas supplies as one Russian vehicle for exerting power, it does not provide a complex assessment of individual countries in the region and also omits any examination of the oil sector. Other area-specific publications include *The Impact of the Russia–Ukraine Gas Crisis in South Eastern Europe* by Aleksandar Kovacevic (2009) and *The Russo-Ukrainian Gas Dispute of January 2009: A Comprehensive Assessment* by Simon Pirani, Jonathan Stern, and Katja Yafimava (2009). These are two of several publications released in the aftermath of the infamous crisis of 2009, a defining event which left a large number of Eastern European countries shivering on exceptionally cold January days. Albeit gas-related, the crisis did draw the attention of the public and scholars to the issue of energy security and gave rise to a number of articles and books. One of the best-known books focused on Russian foreign policy in relation to the concept of energy 'weaponization' is by Anita Orbán (Orbán, 2008).

However, all these publications either are limited in geographical scope or omit the oil industry. Additionally, with a few exceptions, the related literature surprisingly omits the region most affected by the aforementioned crisis—South-Eastern Europe (SEE). When speaking specifically about South-Eastern Europe, one can find publications focused on the entire region (Weithmann, 1996), on individual states (Bešić & Spasojević, 2018; Brkić, 2009; Çukaj, 2015; Jirušek, Vlček, & Henderson, 2017; Maricic, Danilovic, & Lekovic, 2012; Maricic, Danilovic, Lekovic, & Crnogorac, 2018; Tchalakov & Mitev, 2019; Vatansever, 2005; Vlček & Jirušek, 2018 etc.), or, from the other alternative perspective, on Russia and its ties to the region (Ekinci, 2013; Headley, 2008). A specific subgroup consists of works focused on individual Russian companies active in the post-communist countries or elsewhere abroad (Liuhto, 2006; Zashev, 2006) or energy infrastructure development (Mišík & Nosko, 2017). Two books have been published recently that cover Russian economic interests in Central and Eastern Europe, the first edited by Ana Otilia Nuțu and Sorin Ioniță (2017) and the second edited by Ognian Shentov, Ruslan Stefanov, and Martin Vladimirov (2019). Although both books are impressive in terms of detail and geographical coverage, the selection of SEE countries is nevertheless inadequate to derive a complete picture of the region. Except for Romania, which we also include, the first book does not examine the SEE countries at all; the second adds some countries of South-Eastern Europe, namely, Bosnia and Herzegovina, Bulgaria, Macedonia, Serbia, and Montenegro.

As can be seen from the above listing, although undoubtedly of value all of these publications are limited to some extent and have ultimately neglected part of the picture. What is even more surprising is the limited amount of literature that focuses on the contemporary geopolitical and energy-related situation in South-Eastern Europe. This is especially striking given the region's complicated history (especially the Western Balkans) and geopolitical importance. From the energy point of view, the region is also important for potential future energy supplies to other parts of Europe. It is also worth noting that although quite a few titles were published some ten years ago, particularly around the gas crisis of 2009, the pool of more recent titles is considerably smaller. Given the continued prominence of the issue, the authors have responded with this volume to the need for a more recent take on the topic.

To sum up, although there are numerous publications that deal with the role of energy supplies in Russian foreign policy, Russia's relationship to specific regions has by large been neglected. Nor has there been much in the way of a combined, case-based approach to South-Eastern Europe that focuses not only on the energy sphere but also on related policy-shaping factors. Generally speaking, as demonstrated above, most work on the topic of this book is either too broad or too focused to accommodate the nuances of the region, or it neglects other aspects of the issue or tends to simplify the matter to the point of inaccuracy. And finally, the entire region has for the most part been pushed to the sidelines of the discussion. Even in dealing with oil and gas, attention has been focused instead on prominent producers and consumers. In this book, we aim to provide the analysis we believe the region of South-Eastern Europe deserves in light of its noteworthy past and present as well as its significant future role in supplying energy to Europe.

Sources

Aalto, P. (Ed.). (2007). *The EU-Russian Energy Dialogue. Europe's Future Energy Security*. London: Routledge. https://doi.org/10.4324/9781315558455

Aalto, P. (Ed.). (2012). *Russia's Energy Policies. National, Interregional and Global Levels*. Cheltenham: Edward Elgar Publishing. https://doi.org/10.1111/1478-9302.12016_41

Bahgat, G. (2011). *Energy Security: An Interdisciplinary Approach*. Chichester, West Sussex: Wiley.

Balmaceda, M. (2012). *Energy Dependency, Politics and Corruption in the Former Soviet Union*. Routledge.

Balmaceda, M. (2015). *Politics of Energy Dependency: Ukraine, Belarus, and Lithuania Between Domestic Oligarchs and Russian Pressure*. Toronto: University of Toronto Press.

Balzer, H. (2005). The Putin Thesis and Russian Energy Policy. *Post-Soviet Affairs, 21*(3), 210–225. https://doi.org/10.2747/1060-586X.21.3.210

Bechev, D. (2017). *Rival Power: Russia's Influence in Southeast Europe*. New Haven: Yale University Press.

Bešić, M., & Spasojević, D. (2018). Montenegro, NATO and the Divided Society. *Communist and Post-Communist Studies, 51*, 139–150. https://doi.org/10.1016/j.postcomstud.2018.04.006

Bilgin, M. (2011). Energy Security and Russia's Gas Strategy: The Symbiotic Relationship Between the State and Firms. *Communist and Post-Communist Studies, 44*, 119–127. https://doi.org/10.1016/j.postcomstud.2011.04.002

Brkić, D. (2009). Serbian Gas Sector in the Spotlight of Oil and Gas Agreement with Russia. *Energy Policy, 37*, 1925–1938. https://doi.org/10.1016/j.enpol.2009.01.031

Colombano, A., & Colombano, R. (2015). *Oil & Gas Company Analysis: Upstream, Midstream and Downstream*. CreateSpace Independent Publishing Platform.

Considine, J., & Kerr, W. (2002). *The Russian Oil Economy*. Cheltenham: Edward Elgar.

Çukaj, I. (2015). Economic Security as Priority, Energy Security, Advantage of Western Balkans and Albania. *European Journal of Business, Economics and Accountancy, 4*(1), 8–31.

Dannreuther, R. (2016). EU-Russia Energy Relations in Context. *Geopolitics, 21*(4), 913–921. https://doi.org/10.1080/14650045.2016.1222521

Dawisha, K. (2015). *Putin's Kleptocracy: Who Owns Russia?* New York: Simon & Schuster.

de Graaf, T. V., & Colgan, J. D. (2017). Russian Gas Games or Well-oiled Conflict? Energy Security and the 2014 Ukraine Crisis. *Energy Research & Social Science, 24*, 59–64. https://doi.org/10.1016/j.erss.2016.12.018

Deák, A. (2014). *Emerging Rival or Dynamic Partner? The EU and Russia*. In P. Balázs (Ed.), *Europe's Position in the New World Order* (pp. 117–135). Budapest: Center for EU Enlargement Studies of the Central European University.

Dellecker, A., & Gomart, T. (2011). *Russian Energy Security and Foreign Policy*. Routledge.

Downey, M. (2009). *Oil 101*. Wooden Table Press.

Dreger, C., Kholodilin, K. A., Ulbricht, D., & Fidrmuc, J. (2016). Between the Hammer and the Anvil: The Impact of Economic Sanctions and Oil Prices on Russia's Ruble. *Journal of Comparative Economics, 44*, 295–308. https://doi.org/10.1016/j.jce.2015.12.010

Ekinci, D. (2013). *Russia and the Balkans After the Cold War*. Libertas.

Gilpin, R. (2001). *Global Political Economy: Understanding the International Economic Order*. Princeton: Princeton University Press.

Goldthau, A. (2008). Rhetoric Versus Reality: Russian Threats to European Energy Supply. *Energy Policy, 36*, 686–692. https://doi.org/10.1016/j.enpol.2007.10.012

Goldthau, A., & Boersma, T. (2014). The 2014 Ukraine-Russia Crisis: Implications for Energy Markets and Scholarship. *Energy Research & Social Science, 3*, 13–15. https://doi.org/10.1016/j.erss.2014.05.001

Gusev, A., & Westphal, K. (2015). *Russian Energy Policies Revisited*. Research Paper No. 8. Berlin: Stiftung Wissenschaft und Politik. Retrieved from https://www.swp-berlin.org/fileadmin/contents/products/research_papers/2015RP08_gsv_wep.pdf

Hamilton, M. S. (2012). *Energy Policy Analysis: A Conceptual Framework.* Armonk, NY: Routledge.

Harsem, Ø., & Claes, D. H. (2013). The Interdependence of European–Russian Energy Relations. *Energy Policy, 59,* 784–791. https://doi.org/10.1016/j.enpol.2013.04.035

Headley, J. (2008). *Russia and the Balkans: Foreign Policy from Yeltsin to Putin.* Columbia University Press.

Henderson, J. (2015a, April). *Key Determinants for the Future of Russian Oil Production and Exports.* Retrieved from https://www.oxfordenergy.org/wpcms/wp-content/uploads/2015/04/WPM-58.pdf

Henderson, J. (2015b). Russia's Changing Gas Relationship with Europe. *Russian Analytical Digest, 163,* 2–6.

Henderson, J., & Pirani, S. (2014). *The Russian Gas Matrix: How Markets Are Driving Change.* New York: Oxford University Press.

Hill, F. (2004). *Energy Empire: Oil, Gas, and Russia's Revival.* London: The Foreign Policy Centre.

Hill, F., & Gaddy, C. G. (2015). *Mr. Putin: Operative in the Kremlin.* Washington, DC: The Brookings Institution.

Högselius, P. (2013). *Red Gas: Russia and the Origins of European Energy Dependence.* New York: Palgrave Macmillan.

Inkpen, A. C., & Moffett, M. H. (2011). *The Global Oil & Gas Industry: Management, Strategy & Finance.* Tulsa: PennWell.

Jirušek, M., & Kuchyňková, P. (2018). The Conduct of Gazprom in Central and Eastern Europe: A Tool of the Kremlin, or Just an Adaptable Player? *East European Politics and Societies and Cultures, 32*(4), 818–844. https://doi.org/10.1177/0888325417745128

Jirušek, M., Vlček, T., & Henderson, J. (2017). Russia's Energy Relations in Southeastern Europe: An Analysis of Motives in Bulgaria and Greece. *Post-Soviet Affairs, 33*(5), 335–355. https://doi.org/10.1080/1060586X.2017.1341256

Kalicki, J. H., & Goldwyn, D. L. (2013). *Energy and Security: Strategies for a World in Transition* (2nd ed.). Washington, DC: Johns Hopkins University Press.

Kaplan, R. D. (2012). *The Revenge of Geography: What the Map Tells Us About Coming Conflicts and the Battle Against Fate.* New York: Random House.

Kapustin, N. O., & Grushevenko, D. A. (2018). Exploring the Implications of Russian Energy Strategy Project for Oil Refining Sector. *Energy Policy, 117,* 198–207. https://doi.org/10.1016/j.enpol.2018.03.005

Kirchick, J. (2017). *The End of Europe: Dictators, Demagogues, and the Coming Dark Age.* New Haven and London: Yale University Press.

Konoplyanik, A. A. (2012). Russian Gas at European Energy Market: Why Adaptation Is Inevitable. *Energy Strategy Reviews, 1*(1), 42–56. https://doi.org/10.1016/j.esr.2012.02.001

Kovacevic, A. (2009, March). *The Impact of the Russia–Ukraine Gas Crisis in South Eastern Europe*. Retrieved August 25, 2016, from Oxford Institute for Energy Studies: https://www.oxfordenergy.org/wpcms/wp-content/uploads/2010/11/NG29-TheImpactoftheRussiaUkrainianCrisisinSouthEasternEurope-AleksandarKovacevic-2009.pdf

Kropatcheva, E. (2012). Russian Foreign Policy in the Realm of European Security Through the Lens of Neoclassical Realism. *Journal of Eurasian Studies, 3*, 30–40. https://doi.org/10.1016/j.euras.2011.10.004

Liuhto, K. (2006). *Expansion or Exodus: Why Do Russian Corporations Invest Abroad?* Routledge.

Lucas, E. (2014). *The New Cold War: Putin's Russia and the Threat to the West*. St. Martin's Griffin (3rd revised ed.).

Luft, G., & Korin, A. (2009). *Energy Security Challenges for the 21st Century: A Reference Handbook*. Santa Barbara: Praeger Security International.

Mankoff, J. (2009). *Russian Foreign Policy: The Return of Great Power Politics*. Plymouth: Rowman & Littlefield Publishers, Inc.

Maricic, V. K., Danilovic, D., & Lekovic, B. (2012). Serbian Oil Sector: A New Energy Policy Regulatory Framework and Development Strategies. *Energy Policy, 51*, 312–322. https://doi.org/10.1016/j.enpol.2012.08.021

Maricic, V. K., Danilovic, D., Lekovic, B., & Crnogorac, M. (2018). Energy Policy Reforms in the Serbian Oil Sector: An Update. *Energy Policy, 113*, 348–355. https://doi.org/10.1016/j.enpol.2017.11.011

Marshall, T. (2015). *Prisoners of Geography: Ten Maps That Explain Everything About the World*. London: Elliott and Thompson Limited.

Mišík, M., & Nosko, A. (2017). The Eastring Gas Pipeline in the Context of the Central and Eastern European Gas Supply Challenge. *Nature Energy, 2*, 844–848. https://doi.org/10.1038/s41560-017-0019-6

Mišík, M., & Prachárová, V. (2016). Before 'Independence' Arrived: Interdependence in Energy Relations Between Lithuania and Russia. *Geopolitics, 21*(3), 579–604. https://doi.org/10.1080/14650045.2015.1113402

Monaghan, A. (2007). Russia's Energy Diplomacy: A Political Idea Lacking a Strategy? *Southeast European and Black Sea Studies, 7*(2), 275–288. https://doi.org/10.1080/14683850701402201

Moran, D., & Russell, J. A. (2008). *Energy Security and Global Politics: The Militarization of Resource Management*. New York: Routledge.

Morse, E., & Richard, J. (2002). The Battle for Energy Dominance. *Foreign Affairs, 81*(2), 16–31.

Myers, S. L. (2015). *The New Tsar: The Rise and Reign of Vladimir Putin*. New York: Alfred A. Knopf.

Newnham, R. (2011). Oil, Carrots, and Sticks: Russia's Energy Resources as a Foreign Policy Tool. *Journal of Eurasian Studies, 2*, 134–143. https://doi.org/10.1016/j.euras.2011.03.004

Nuţu, A. O., & Ioniţă, S. (Eds.). (2017). *Energy, Russian Influence, and Democratic Backsliding in Central and Eastern Europe: A Comparative Assessment and Case Studies from Belarus, Ukraine, Moldova, Hungary, Romania*. Bucharest: Expert Forum. https://expertforum.ro/en/files/2017/05/Final-countries-report-coperta.pdf

Nygren, B. (2007). *The Rebuilding of Greater Russia: Putin's Foreign Policy Towards the CIS Countries*. London: Routledge.

Orbán, A. (2008). *Power, Energy, and the New Russian Imperialism*. Westport: Praeger Publishing.

Orttung, R., & Overland, I. (2010). A Limited Toolbox: Explaining the Constraints on Russia's Foreign Energy Policy. *Journal of Eurasian Studies, 2*, 74–85. https://doi.org/10.1016/j.euras.2010.10.006

Osička, J., & Ocelík, P. (2017, March). Natural Gas Infrastructure and Supply Patterns in Eastern Europe: Trends and Policies. *Energy Sources Part B Economics, Planning and Policy, 12*(4). https://doi.org/10.1080/15567249.2015.1136971

Parra, F. (2010). *Oil Politics: A Modern History of Petroleum*. London: I.B. Tauris.

Perović, J., Orttung, R. W., & Wenger, A. (2009). *Russian Energy Power and Foreign Relations: Implications for Conflict and Cooperation*. London: Routledge.

Pirani, S., Stern, J., & Yafimava, K. (2009, February). *The Russo-Ukrainian Gas Dispute of January 2009: A Comprehensive Assessment*. Retrieved April 10, 2015, from Oxford Institute for Energy Studies: http://storage.globalcitizen.net/data/topic/knowledge/uploads/20110630223130533.pdf

Richter, P. M., & Holz, F. (2015). All Quiet on the Eastern Front? Disruption Scenarios of Russian Natural Gas Supply to Europe. *Energy Policy, 80*, 177–189. https://doi.org/10.1016/j.enpol.2015.01.024

Romanova, T. (2014). Russian Energy in the EU Market: Bolstered Institutions and Their Effects. *Energy Policy, 74*, 44–53. https://doi.org/10.1016/j.enpol.2014.07.019

Salminen, V. (2015). *Putin: Nezkreslená zpráva o mocném muži a jeho zemi* [Putin: An Undistorted Message About the Mighty Man and His Country]. Prague: Daranus.

Schmidt-Felzmann, A. (2011). EU Member States' Energy Relations with Russia: Conflicting Approaches to Securing Natural Gas Supplies. *Geopolitics, 16*(3), 574–599.

Shentov, O., Stefanov, R., & Vladimirov, M. (2019). *The Russian Economic Grip on Central and Eastern Europe*. Abingdon: Routledge. ISBN 978-0-8153-6342-2.

Stegen, K. S. (2011). Deconstructing the "Energy Weapon": Russia's Threat to Europe as Case Study. *Energy Policy, 39*, 6505–6513. https://doi.org/10.1016/j.enpol.2011.07.051

Stern, J. (2005). *The Future of Russian Gas and Gazprom*. Oxford: Oxford University Press.

Stulberg, A. N. (2015). Out of Gas?: Russia, Ukraine, Europe, and the Changing Geopolitics of Natural Gas. *Problems of Post-Communism, 62*, 112–130. https://doi.org/10.1080/10758216.2015.1010914

Svarin, D. (2016). The Construction of 'Geopolitical Spaces' in Russian Foreign Policy Discourse Before and After the Ukraine Crisis. *Journal of Eurasian Studies, 7*, 129–140. https://doi.org/10.1016/j.euras.2015.11.002

Taylor, B. D. (2018). *The Code of Putinism*. New York: Oxford University Press.

Tchalakov, I., & Mitev, T. (2019). Energy Dependence Behind the Iron Curtain: The Bulgarian Experience. *Energy Policy, 126*, 47–56. https://doi.org/10.1016/j.enpol.2018.11.008

Tsygankov, A. (2016). *Russia's Foreign Policy: Change and Continuity in National Identity* (4th ed.). Lanham: Rowman & Littlefield.

Tsygankov, A. P. (2010). *Russia's Foreign Policy: Change and Continuity in National Identity*. Plymouth: Rowman & Littlefield Publishers, Inc.

Vatansever, A. (2005). *Bulgaria: Russian Involvement in Bulgaria's Oil Sector*. Kogan Page Business Books.

Vatansever, A. (2017). Is Russia Building Too Many Pipelines? Explaining Russia's Oil and Gas Export Strategy. *Energy Policy, 108*, 1–11. https://doi.org/10.1016/j.enpol.2017.05.038

Victor, D. G., Jaffe, A. M., & Hayes, M. H. (2008). *Natural Gas and Geopolitics: From 1970 to 2040*. Cambridge: Cambridge University Press.

Vlček, T., & Jirušek, M. (2018). The Hydrocarbons Sector in Albania: Short-Term Challenges and Long- Term Opportunities. *Mediterranean Quarterly, 29*(1), 96–119. https://doi.org/10.1215/10474552-4397358

Weithmann, M. W. (1996). *Balkán – 2000 let mezi východem a západem*. Český Těšín: Vyšehrad.

Woehrel, S. (2012). Russian Energy Policy Toward Neighboring Countries. *Current Politics and Economics of Europe, 23*(3/4), 403–433.

Yergin, D. (2008). *The Prize: The Epic Quest for Oil, Money & Power*. New York: Free Press.

Yergin, D. (2012). *The Quest: Energy, Security, and the Remaking of the Modern World*. New York: Penguin Books.

Youngs, R. (2009). *Energy Security: Europe's New Foreign Policy Challenge* (Routledge Advances in European Politics). Routledge.

Zashev, P. (2006, September). Russian Companies in the Forthcoming EU Member States: A Case of Lukoil in Bulgaria. *Journal of East-West Business, 11*(3–4), 109–128.

CHAPTER 3

Research Design

3.1 THEORETICAL FRAMEWORK[1]

The research process underlying this book targeted the following question: 'Do Russian state-owned energy companies in the oil sector behave in South-Eastern European countries as tools of the Russian state, and do they serve as vehicles of Russian foreign policy?' The purpose of this chapter is to explain the methodology behind the research used to answer this question. First, the research area, principles, and logic are described. Then, the chief tools employed in the research are set out and an outline of the research process itself is provided. Finally, the ideal type model used to analyse individual countries/cases is presented along with the pertinent theoretical background.

Conventionally, social science methodology distinguishes between qualitative and quantitative approaches, and scholars sometimes portray the differences between them as being so wide that they are incompatible with each other. The authors of this text, however, view qualitative and quantitative research as parts of a single continuum where a particular

---

[1] This chapter partially builds on and develops books titled *Politicization in the Natural Gas Sector in South-Eastern Europe: Thing of the Past or Vivid Present?* (Jirušek, 2017), *Energy Security in Central and Eastern Europe and the Operations of Russian State-Owned Energy Enterprises* (Jirušek et al., 2015), both published by Masaryk University Press, and a paper titled 'Russia's Energy Relations in Southeastern Europe: An Analysis of Motives in Bulgaria and Greece' (Jirušek, Vlček, & Henderson, 2017), published in the Taylor & Francis journal *Post-Soviet Affairs* (www.tandfonline.com).

© The Author(s) 2019
T. Vlček, M. Jirušek, *Russian Oil Enterprises in Europe*,
https://doi.org/10.1007/978-3-030-19839-8_3

study may fall closer to one end than the other. Nor should the distinction between the two ever be the defining principle for research. Rather, the researcher should focus on the research goal and choose a methodology strictly on that basis. Here, reference may be made to the well-known work of Gary King, Sidney Verba, and Robert O. Keohane, *Designing Social Inquiry: Scientific Inference in Qualitative Research*, in which the authors attempt to bridge the gap that has emerged between proponents of qualitative and quantitative research (King, Verba, & Keohane, 1994). They note that the distinction between qualitative and quantitative measures is more a question of research 'style' and level of abstraction. In effect, they insist that most research is neither exclusively qualitative nor exclusively quantitative; rather it combines both, depending upon the level of analysis (King et al., 1994, pp. 3–5). The research undertaken here may not be strictly labelled as either purely qualitative or purely quantitative in nature. At the most general level, however, given the research process, the methodology employed, and the way the data are treated, it may be perceived as predominantly qualitative. This is perhaps most visible in the fact that the case study approach is used to examine individual countries. The deep understanding this brings and the clear-cut mapping it provides of individual countries are also characteristic of that approach[2] (Creswell, 2009, p. 164). The way the findings are communicated also reveals the predominantly qualitative nature of the research, particularly as regards its use of narrative (ibid., p. 186).

3.2   Logic of the Research, Research Area, and Basic Principles

As suggested above, the goal of the research was to determine whether Russia's conduct in the countries under analysis follows a particular logic, and whether this logic seems to support charges that Russia abuses energy supplies as political leverage. The authors took an established theoretical approach to energy policy, the so-called strategic approach to energy policy (see Sect. 3.5), which is suited to evaluating the type of accusations brought against Russia, and sought to find out whether the actions of Russian companies and Russian officials in fact correspond with that approach. In this sense, the research utilized an inductive, theory-driven

---

[2] The data-gathering techniques used and reference to multiple sources are also indicative of qualitative research as described in Creswell (2009, p. 164).

logic based on in-depth case studies that include field research in the individual countries, thereby allowing general conclusions to be inferred from specific evidence. The authors resorted to this bottom-up research logic for two main reasons. First, the goal was to get a realistic notion of the actual tactics applied in individual countries that an image of the overall strategy could then be constructed upon. This would be impossible if the research logic was inverted and a top-down/deductive approach used instead. It would fail to describe the objective reality of the individual cases, something which is crucial for this work. Second, a realistic description of the overall Russian strategy would thereby become difficult to acquire (if indeed a coherent strategy has been formulated). State and company officials would hardly be eager to share the details, and expecting sincere answers would be naive. In addition, a general strategy would not provide concrete clues and the much-needed guidance for conducting the in-depth case studies. The research would thus risk coming to inaccurate conclusions.

Furthermore, the research area must be defined. This was chosen to be the oil sectors of the countries of South-Eastern Europe (SEE).[3] As noted above, case studies are used to examine individual countries in the region in keeping with the predominantly qualitative nature of the research.[4] The particular cases under scrutiny are defined in the sense of John Gerring (2007), who says that a case is essentially an example of a particular phenomenon broadly delimited spatially and (usually) also temporally (Gerring, 2007, p. 19). In this context, cases are defined as states within the South-Eastern European region examined within a timeframe stretching from the fall of communism to the present day. Although the authors accept Gerring's definition of the case, the study outcomes are defined not by inferring population characteristics, but rather as the sum of findings of in-depth (intrinsic) case studies. This is because the sample covers the whole population: all the countries of South-Eastern Europe with energy-related ties to Russia attributable to their history, political background, and geographic location. The cases are therefore as follows (in alphabetical order): Albania, Bosnia and Herzegovina, Bulgaria, Croatia, Greece,

---

[3] The book also includes a separate chapter providing an overview of the natural gas sector in the region, but this is included for the sake of comparison with the main focus of interest the oil sector.

[4] Although Gerring notes that assigning the case study to qualitative methods is a '(...) methodological affinity, not a definitional entailment' (Gerring, 2007, pp. 10–11), the authors adhere to this prevailing classification.

Kosovo, Macedonia, Montenegro, Romania, Serbia, and Slovenia. In the sense of Gerring's typology, the publication bears the features of a cross-sectional research (Gerring, 2007, p. 28).

The first step in the research process was to create the data collection for each case study. The authors focused on openly available sources, gathering data from official documents and statistics on the energy sectors of the countries under scrutiny and published by universally recognized institutions such as the Energy Community, the European Commission, the International Energy Agency, and others. Supporting and complementary information needed to understand the cases in their full complexity (e.g. formative events or specific factors influencing individual cases) were collected from newspapers and specialized websites. Sources of this type were varied in nature, ranging from the informative to the analytical or investigative. Crucial to the research process were semi-structured interviews and thorough field studies carried out in the SEE countries. For purposes of the interviews, the authors selected local analysts, experts in the field, and journalists as well as, to some extent, governmental representatives.[5] The interviews, conducted between spring 2014 and spring 2017, proved a valuable source of information for filling in the blanks or where there was insufficient overall coverage of the subject matter. When a point was ambiguous, the authors verified the information from more than one source, often combining openly available sources with interviews. Sources were double-checked as a rule throughout the research process when seeking out specific information. Independent sources were then usually combined to give the least distorted picture of reality possible.

The timespan of the research stretches from the dissolution of the Soviet Bloc and the emancipation of the countries in the region under analysis to the present. The deeper roots of the development of the oil sector, stretching beyond the fall of communism into the more distant past, were also taken into account when necessary, complementing the research. The research was conducted between November 2013 and July 2018, with the end date set at July 31.

---

[5] Governmental representatives (i.e. mainly representatives of the pertinent ministries) were chosen carefully and used as sources of general information to compensate for their potential bias.

3.3   CASE STUDIES

As noted, the focus will be on the following states: Albania, Bosnia and Herzegovina, Bulgaria, Croatia, Greece, Kosovo, Macedonia, Montenegro, Romania, Serbia, and Slovenia. The SEE states were selected because they form a geographically coherent region, one in which the majority of states (with the single exception of Greece[6]) share a common history of development in their energy sectors, as well as a common experience of authoritarian regimes and membership in the so-called Eastern Bloc. This experience as satellites subordinated to the Soviet Union is of overarching importance, having influenced the whole of postwar development in the region. The list of these countries consists of relatively underdeveloped energy sectors, where conventional traditional fuels still dominate, not only in terms of installed capacity but political and strategic support as well. Oil, as a commodity, is easy to transport, relatively simple to process in refineries, and easy to use (the vast majority in transportation). Hence, oil is one of the principal fuels in the SEE region. Very few countries in the region have relevant domestic resources of crude oil (more than 25% of domestic consumption). The list of these countries is very short—Romania and Serbia. This major limitation in domestic reserves makes nearly all the selected countries dependent on crude oil imports. This reality, together with geographic coherence, makes the SEE region a logical group of cases to analyse.

The core of the research thus comprises case studies, each dedicated to one of the states under scrutiny. To assess the Russian state-owned oil companies' behaviour, the authors have constructed an ideal type model of strategically motivated behaviour consisting of a set of features and evidence as to whether they have manifested in reality (i.e. a set of indicators). Each indicator was assessed for each individual case to determine the extent to which Russian state-owned oil companies and their subsidiaries have engaged in that particular strategically motivated behaviour.

Given the emphasis on individual cases, the research was conceived as a set of individual case studies (see e.g. Gerring, 2007, p. 20). The use of a set of individual case studies is especially appropriate from the standpoint of the research goal for each case study, which is to obtain a deep understanding of the case at hand. The focus is on the behaviour of the subject under exami-

---

[6] However, Greece is also partially supplied by the infrastructure that generally feeds the other states under scrutiny. The country has also had a somewhat special relation with Russia and belongs geographically to the region, which is poised to take on importance for transit infrastructure that is either in planning or already under construction.

nation (Russian state-owned oil companies) in various environments. Given the diversity of the internal environments of the individual states, understanding must be both broad and deep. Rather than generalizing, then, the aim has been to let the outcomes of individual cases speak for themselves. This bottom-up procedure proved valuable, as after comparing of the outcomes of individual cases studies, common characteristics were derived that create a more comprehensive picture of the general strategy taken by Russian companies in the region. A further advantage of this choice of design is its versatility. A research design comprising a set of individual cases treated as independent may be expanded or restricted based upon the population or region the author must investigate. This makes the design amenable to further use. Analogously, the design may be used to conduct a single case study aimed at investigating a single state or a cluster of several states.

The research is theory-driven, here represented by an ideal type model of strategic behaviour in the oil sector (see below) which was constructed beforehand and used as a model behaviour to which the conduct of Russian companies was compared in individual cases. Indicators were assessed to determine the extent to which Russian state-owned oil companies or the Russian government engaged in the behaviour in question. Hence, the research upon which this publication is based utilized the principles of the so-called idiographic, theory-guided case study as described, for instance, by Jack S. Levy (Levy, 2008). Levy characterizes as idiographic those case studies that describe, explain, or interpret a particular case (Levy, 2008, p. 3), which is essentially what case studies contained in this book do. In addition to Levy, this type of study, guided by a previously defined theory or concept, is described by Robert E. Stake (2006, pp. 2–16) and John S. Odell (2004, pp. 58–61), who use the term 'disciplined interpretative study'. In their understanding, the disciplined interpretative study uses a theory-driven approach, where a previously created model (in this instance, strategic behaviour in the oil sector, see below) is applied to specific cases (here the SEE countries) (Kořan, 2008, pp. 34–39; Odell, 2004, pp. 58–61; Stake, 2006, pp. 2–16).

3.4   THE RESEARCH PROCESS

To refine the large amount of data, basic data mining principles have been utilized as used in content analysis, for example (see for instance Holsti, 1969 or Krippendorf, 2013). For the initial step involving construction of the profiles of oil sectors in individual countries, the authors selected relevant documents and other sources and employed the basic methods of open coding (as used in the method of content analysis noted) to get oriented in what was a huge quantity of information. This part of the research process included categorizing information from international organizations (e.g. the Energy Community, International Energy Agency, European Commission) and specialized agencies belonging to the individual governments (mostly ministries or regulatory authorities) to determine relevance with regard to the oil sector of each country under scrutiny. For this purpose, coding units (previously defined pieces of information with the needed information value) were defined (i.e. unitized) as units bearing information on the oil sector of the country in question (thematic distinction). The coding units were recognized within the texts with the help of surrounding information (i.e. contextual units) (Krippendorf, 2013, pp. 99–103).

Once this initial step was complete, the focus was on locating and filling in pieces of information that had not been found during the first step. To do so, the same methods were utilized once again, this time focused specifically on the missing bit of missing information. Coding units were therefore defined (unitized) once again, this time as units bearing that particular bit of information that had been missing. Once again, surrounding contextual units helped in locating the salient information. This process of reusing the same technique, with a more precise definition of what to look for, aided in finding what had been omitted. In the second step, sources of information were broadened to include newspaper articles and reports by analytical and investigative institutions.

The third step involved utilizing the same procedure as in steps one and two, this time focusing on each indicator (see below). Coding units were thereby unitized with regard to each indicator. Then, information sources (combining those mentioned above) were refined to locate related information. Once again, surrounding (contextual) information was used as a guideline.

In cases where some information was still missing even after the steps noted above had been performed, or when information was uncertain or needed verification, the authors conducted semi-structured interviews to fill in the gaps or provide verification. These interviews differed from case to case, based upon what information was missing in each case study. The choice to use semi-structured interviews proved correct, since it provided the authors (i.e. the interviewers) with clear guidance throughout the interview based upon the indicators common to all cases; at the same time it allowed topical trajectories to be tracked where necessary (Brinkmann, 2013, p. 21). The semi-structured interview thus enabled adequate guidance balanced with freedom. In addition, interviewees were selected on the basis of their background and experience to further increase the reliability of the research. Interviewees were primarily local analysts and academics, field-related experts, journalists, and in some instances governmental representatives.[7] Most insisted upon anonymity because of job pressure or the complexities of getting their responses authorized. In view of this, the authors decided to code the respondents' names. In each case study, therefore, respondents are coded simply 'Interviewee', appended with a number to differentiate them. Their identity is, of course, known to the authors.

The data-gathering procedure described above shows the emphasis the authors placed on the verification and triangulation of data. As a general rule, the information underlying the indicators (as the cornerstone of the research and key components for deriving outcomes on the behaviour of Russian oil companies) was always verified against at least two independent sources. Where such verification was impossible, the authors made sure to note this in the text and not derive a definite outcome for such a finding.

Each case study begins with a description of the country's oil sector, with particular regard to its relations with Russia. Features related to other energy supplies specific to the country are also briefly described so that the reader may develop a comprehensive picture of the country. Highlighted are the characteristics of supply infrastructure, contracts, the structure of companies active in the market and their relation to the state, the potential involvement of Russian capital and/or politicians on either side, the history of mutual relations—including any relevant disputes—and so on. The final portion of each study comprises reflections on the indicators (for a list of these, see the subsection below). Here, most of the findings treated in detail in the prior section of the case study are categorized and summarized to offer a comprehensive overview of the situation. For the sake of completeness, a table summarizing the indicators over all cases is included at the end of the book.

---

[7] Usually, governmental representatives were not the first choice, the reason being potential bias. Therefore, they were usually interviewed for purposes of gathering general information on the sector where this could not be had otherwise.

As noted above, the last part of the book is dedicated to an overview of the natural gas sector of the region. This chapter was included mainly for two reasons. First, it is usually Gazprom, the Russian state-owned gas company, which is accused of serving as the Russian government's foreign policy lever. Second, the oil and gas industries have been largely intertwined for most of their history. Although over the past two decades the two sectors have gone in different directions when it comes to utilization and, mainly, transport and marketing, they do still bear common features. The intention behind including the overview is to compare the situation in both sectors so as to capture similarities between the two. This provides a better understanding of the behavioural patterns of Russian companies in the energy sector and any potential ties to the Kremlin. This portion of the book is based on research conducted earlier by the authors based on a similar methodology adapted to the needs of the natural gas sector (Jirušek, 2017; Jirušek et al., 2015; Jirušek & Kuchyňková, 2018; Jirušek, Vlček, & Henderson, 2017).[8]

3.5 Theoretical Basis of the Model for Assessing the Oil Sector

To examine the oil sector of the states under discussion, the authors constructed an ideal type model defined by a set of features and evidence of their manifestation in reality (the indicators) sought for in each case. This model is based on the strategic approach to energy policy and its theoretical foundations (see for instance Klare, 2005, 2009a; Leverett, 2009; Luft & Korin, 2009; Moran, 2009). The assumptions that underlie the approach stem from the theoretical basis provided by classical realism, neorealism, and neoclassical realism. All these are described in the section that follows, but a simple rendering is that the strategic approach centres on behaviour that does not result in capitalization in the foreseeable future, and that generally rejects economic logic as the main determining factor behind energy policy. Instead, it is characterized by the (mis)use of energy commodities as a tool, often with the aim of achieving certain foreign policy goals.

---

[8] Both methodologies share the same theoretical foundations, principles, and research logic, and treat the data in like ways. Their differences lie in differences between the oil and gas sectors as such, mainly in terms of transport, distribution, and marketing. In order to address them, different indicators were used for the various distinct areas.

The strategic approach to energy policy and security as a basis for the ideal type model utilized in this research was chosen to address the frequent accusations, noted above, that Russia misuses energy supplies and energy-related deals to exert pressure on its customers. Such behaviour corresponds to the characteristics of the strategic approach to energy policy and security, which represents one of the two main approaches to energy policy in general. Essentially, the cleavage follows the division between the two major paradigms in the field of international relations, realism and liberalism, and their variations. Similarly, there are two main approaches to energy policy in the field of energy security (Luft & Korin, 2009) that basically follow their theoretical origins—the strategic approach, resting on the assumptions of realism, and the market-oriented approach, based upon liberal theories. In this way, the book is related to one of the overarching issues within the field.

Realists place the state, power, and survival at the centre of attention and insist on the conflictual nature of international relations, asserting that individual actors try to increase their own potential at the expense of others (Jackson & Sorensen, 2007, p. 60). Theorists of liberalism, by contrast, maintain that states as well as individuals may coexist and cooperate and benefit from the collaborative environment thus created (Jackson & Sorensen, 2007, pp. 98–99). Within this perspective, the market and market-related principles take on added importance. Likewise, the sphere of energy policy and security where, as noted above, there is also a split between those who adhere to the realist paradigm and those who adhere to the liberal paradigm. There are, of course, other ways of approaching the situation within the field. But the prominence of these original paradigms in international relations means it is these two approaches that dominate the energy policy and security discourse (see for instance Klare, 2005, 2009a; Leverett, 2009; Luft & Korin, 2009; Moran, 2009). Additionally, the interconnection of both approaches to the sphere of international relations gives them relevance for analysing interstate relations.

These approaches do not, naturally, exist in pure form anywhere in the world. They are opposite endpoints along a single axis, and most policies are located somewhere between the two. The authors have chosen to concentrate on the strategic perspective and constructed the ideal type model as an ideal or standardized strategically oriented behaviour to see the extent to which actions of Russian state-owned oil companies place near this end of the axis. Since the use of energy commodities as tools is a defining feature of the strategic approach to energy policy, the focus is on examining the extent to which Russia practises this behaviour.

Although the goal of the book is to assess whether the accusations that Russian state-owned oil companies are misused as political tools correspond with reality, it is also useful to understand the opposite end of the aforementioned axis. Therefore, to achieve a comprehensive understanding of the issue, a brief overview of the market-oriented approach and a comparison of the two approaches have been added at the end of the following subsection. Examining the contrasts between them helps to clarify individual features of the ideal type model used in this book.

3.5.1 The Realist Tradition of Thought as a Basis for the Ideal Type Model

3.5.1.1 Classical Realism

The realist tradition of thought in international relations is primarily based upon the concept of power. States are unitary actors, superior to all other units in the system, driven by a universal aim of surviving in a hostile environment. Originally allied to this way of thinking was the perception that the state may be treated as a 'black box', a unitary entity whose internal structure and policymaking processes may be neglected. But this perception has undergone substantial revision in later versions of realist theories (see below). Power is the defining principle, and the main objective of all actors is to gain the upper hand over their competitors (Burchill et al., 2005, pp. 30–34). From this standpoint, mutual relations are seen as a zero-sum game, in which one actor's gain is another actor's loss.[9] Military power is seen as crucial but other means of power are also important (Gilpin, 2001, pp. 17–19, 21–24). Among these, it may be economic power which is perceived as the most universal, both because it may be converted into military power and because it is a key determinant of overall state power. The realist tradition further suggests that all activities conducted by or on behalf of the state should be subordinate to its needs. Resources should therefore be used to further the state's prosperity, and companies (typically state-owned) should act with the state's well-being in mind. In this sense, state-owned companies are used as tools to maximize

---

[9] Applied to energy resources, this means that given the scarcity and limited amount of traditional energy resources, if an actor manages to acquire certain energy sources for itself, these are lost for the others and therefore they become weaker.

the economic and thus overall power of the home country. These demands are best met by a state-managed economy rather than one subjected to market-oriented principles. According to Robert Gilpin, although companies usually follow the economic logic of behaviour, their states of origin cannot be separated from the underlying rationale of their conduct (Gilpin, 2001, pp. 17–19, 21–24). When they are owned by a state that views the sector from a strategic perspective, their behaviour follows the state's perception of reality (see below in the section dedicated to neoclassical realism). Gilpin suggests that this is detectable also in the behaviour of privately owned companies, since they operate within and are influenced by an environment created and maintained by the state (Gilpin, 2001, pp. 23–24). Therefore, even for private capital, the state of origin is an influential factor. This is especially true in Russia where, as previously mentioned, state oversight has tightened under Vladimir Putin to the point that virtually no significant energy company would go against the Kremlin's will. This is further fostered by the key figures who have assumed leadership posts in important companies with a strong affinity for the Kremlin. In the cases under examination here, these assumptions are highlighted by the fact that the export of energy sources represents a sizable portion of the Russian economy, and that the SEE countries are often vitally dependent upon Russian supplies of energy commodities and natural resources. Therefore, the state's interest in exporting the goods in question, as well as in maintaining a key position abroad, is obvious.

Under the realist paradigm, a stronger state (i.e. one with greater capacity in terms of natural resources, population, economic output, military strength, etc.) cannot 'resist' the urge to exert its power against weaker actors (states) and is thus likely to (mis)use its dominance by leveraging those sources of power best suited to the situation. As energy commodities are usually vital to a state's economy, they are a natural choice for an actor with control over them (Burchill et al., 2005, pp. 29–34; Jackson & Sorensen, 2007, pp. 60–67). This assumption may be used to explain the misuse of energy supplies in international relations in general and fears of a dependence on supplies of Russian origin in SEE in particular (Table 3.1).

3.5.1.2   Neorealist Theory (Structural Realism)

Neorealism further extended the original assumptions of realism. Unlike classical realism, which emphasizes the role of individual actors, neorealism emphasizes the role of structure. Structure is influenced by the actors but influences them in return. Neorealism is thus often referred to as

Table 3.1 Basic assumptions of classical realism: summary

Based on the concept of power
States as main actors in the system driven by the universal goal to survive in a hostile environment
States are driven to gain superiority over other actors
Relies on the logic of classical geopolitics (e.g. geographical determination influencing views on pipeline policy, transit chokepoints)
Interstate relations are seen as a zero-sum game
Military power is seen as most important
Other means of power are also important; economic power may be perceived as the most universal and may be converted into military power
Energy is seen as a scarce commodity vital to a state's existence
State involvement in the energy sector is essential
Market forces are not seen as reliable; states aim at maintaining control over resources and supply routes
Based on the concept of power
States as main actors in the system driven by the universal goal to survive in a hostile environment
States are driven to gain superiority over other actors
Relies on the logic of classical geopolitics (e.g. geographical determination influencing views on pipeline policy, transit chokepoints)
Interstate relations are seen as a zero-sum game
Military power is seen as most important
Other means of power are also important; economic power may be perceived as the most universal and may be converted into military power
Energy is seen as a scarce commodity vital to a state's existence
State involvement in the energy sector is essential

Source: Burchill et al., 2005, pp. 30–34; Gilpin, 2001, pp. 17–19, 21–24; Jackson & Sorensen, 2007, pp. 60–67

'structural realism'. The structure in question will be either hierarchically ordered or anarchic. In an anarchic system, there are neither roles ascribed to the states within the system nor any overarching authority that rules the system. Instead the actors vie with one another for superiority. If, by contrast, the structure is ordered hierarchically, roles or functions are ascribed on the basis of state characteristics, capacities, or location. In such a setting, some actors become superior or subordinate to others (Burchill et al., 2005, pp. 34–35). The number of centres of power present must also be taken into account. Uni-, bi-, or multipolarity may occur (ibid., pp. 38–39). Although this point is not exclusively tied to energy security, it may be used to explain the way some actors in the system perceive reality.

The implications for energy security may be twofold: first, it is assumed that any state which possesses energy resources is likely to dominate those that do not, as well as subordinate states with lesser capacity. Second, the distribution of energy resources and their use forms the basis for the distribution of roles within the supply chain. Essentially, this means the distribution of roles among producers, transit countries, and consuming states; countries act in keeping with their obligation to transport commodities and where hierarchical relations may occur, even if these are not explicitly formulated.

Within neorealist theory, a division is present between offensive and defensive realism. Offensive realism focuses on the relative gains that may be had over competitors—expansion and short-term gains (Burchill et al., 2005, pp. 43–44; Jackson & Sorensen, 2007, p. 310). Defensive realism aims at conservation of the status quo and maximization of security (i.e. survival, long-term gains) (Burchill et al., 2005, pp. 43–44; Jackson & Sorensen, 2007, p. 306; Waltz, 1979, pp. 79–101). The strategic approach to energy policy and security builds on the assumptions of offensive realism that see state use of energy resources as a way to expand and exert power abroad, even if there is no imminent objective threat to the state's security. The assumptions of offensive realism also provide a basis for using state-owned (energy) companies as foreign policy tools employed in order to achieve relative gains (Burchill et al., 2005, pp. 43–44; Gilpin, 2001, pp. 15–24; Jackson & Sorensen, 2007, pp. 86–88) (Table 3.2).

3.5.1.3 Neoclassical Realism

Neoclassical realism shares the basic assumptions of classical realism but with a greater focus on the individual political actors who drive state policies, and statecraft. It emphasizes the importance of individuals as factors substantially influencing state behaviour within the system. More impor-

Table 3.2 Basic assumptions of neorealism: summary

Emphasizes the role of structure and recognizes the importance of interactions between states
Recognizes the hierarchical order or anarchical order of international relations
Roles/functions are ascribed according to state characteristics (e.g. producer, transit, and consumer states) and position in the system (superiority/subordination)
Emphasizes the importance of relative gains over competitors

Source: Burchill et al., 2005, pp. 34–35, 38–39, 43–44; Gilpin, 2001, pp. 15–24; Jackson & Sorensen, 2007, pp. 86–88, 306, 310; Waltz, 1979, pp. 79–101

tantly, it stresses the personal views of state representatives, and their abilities in political practice. The reactions and behaviour of states are thus strongly linked to the behaviour of state representatives, their values, and their aims. This may explain the discrepancy between the behaviour of Western states, their expectations and reactions, and those of other countries that are culturally and politically different. If a country feels the system is behaving in a hostile manner, it may implement certain countermeasures regardless of objective reality. In the present case, this may explain the differing perceptions of NATO expansion after the Cold War, as well as differing perceptions of the SEE region by Western states versus Russia (Jackson & Sorensen, 2007, pp. 67–74).

3.5.2 Strategic Approach to Energy Policy and Security

The aforementioned theories gave birth to the so-called strategic approach to energy policy and security (see e.g. Klare, 2005, 2009a; Luft & Korin, 2009; Moran, 2009). The strategic approach is not, strictly stated, a complex theory as we understand theories in international relations. It is rather an approach, a general policy or set of features tied to the aforementioned realist paradigm and its variants. It emphasizes the anarchical nature of international relations and power, which is based on material factors including energy sources, which are necessary for the functioning of the economy and the military sector and may also serve as foreign policy tools (Burchill et al., 2005, pp. 29–54). In combination with the theories of neorealism and neoclassical realism, it also accepts the role of structure in the system, insofar as states may form alliances and organizations, but only as a tool of national politics (Waltz, 1979, pp. 79–101). It also recognizes the distribution of roles within the system based on states' roles in producing, transporting, and consuming resources. Elements of classic geopolitics, which are also among the sources of the strategic approach, highlight the importance of geographical advantages and disadvantages associated with energy (these assumptions are reflected in debates over energy autarchy, pipeline policy, and chokepoints; see Klare, 2014). Neoclassical realism also infused the approach with a recognition of the importance of individual policymakers and decision-making processes.

The practical consequences of implementing this approach result in the energy sector being perceived as a strategically important area for state involvement, if the state is to ensure the commodities essential for its survival are available. Such thinking legitimizes direct action by the state and a distinctive role for the energy sector in the economy; analogously, market forces are seen as incapable of securing flows of energy. States' energy policies are thus developed by means of resource nationalism, or resource mercantilism. Producers attempt to strengthen their control over their energy resources; importer countries, on the other hand, work to gain exclusive rights to energy sources, or to strengthen their ties with producers by direct government engagement (Leverett, 2009, p. 214). Energy sources are thus understood not only as a means to be utilized in conflict, but also as a possible cause (Ciuta, 2010, pp. 129–130). Strategic perceptions of energy sources, their gradual exhaustion, and the increasing disputes over these resources are the focus of one of the most prolific proponents of the strategic approach, Michael T. Klare. Klare argues that using the means of government to obtain energy-related resources is comparable to other means of ensuring national security (Klare, 2005, 2009a, 2009b). The relationship between consumers and producers of energy resources is thus that they are both players in a zero-sum game (see for instance Tunsjø, 2010, p. 27). Risks resulting from the disruption of such relations create incentives to diversify the means of transit and source countries, as well as to diversify the consuming country's overall energy mix.

3.5.3   Market-Based Approaches: Opposition to Strategic Measures

For a complete understanding, it is useful to also define the opposing stream of theory. The market-based (or market-oriented) approach is generally a mirror image of the strategic approach, asserting the direct opposite for the majority of features. By defining the market-oriented approach, we can refine the features of the strategic approach by contrasting them to those of the market-oriented approach. Also, we may show the fundamental differences in both approaches stemming from their paradigmatic foundations.

The paradigmatic ground of the market-oriented approach may be found in the writings of authors concerned with neoclassical and neo-institutional economics, and with the liberal theory of international relations. Under this point of view, it is the market and market forces that most effectively allocate energy sources. Rationally informed actors select the optimal strategy to secure energy resources, and governmental influence is seen as a negative, ineffective disruption of this mechanism (see for instance Nordhaus, 2009). In this sense, any emphasis on the security of supply may actually worsen the situation, since it essentially prevents the market from functioning properly (Chester, 2009). Economic factors are more important than those of a political or geopolitical nature; energy commodities are not perceived as particularly unique, and they can and should be seen as standard goods. Efforts by the state to achieve independence lead to disturbances in the system and increase tension. Fears of an anarchistic, hostile international system weaken international institutions and economic cooperation (Ciuta, 2010). The basic premise of the entire approach, the rationality of an actor, strongly questions the necessity for unending confrontation and tends to support cooperation and gains for all participating sides. Fears of resource scarcity expressed by proponents of the strategic approach are addressed by, for example, Morris Adelman. He argues that supplies of energy resources are strictly a function of price: price increases lead to new technologies or shift to alternatives (Adelman, 1973, p. 73; Yergin, 2005, 2006). Also, the use of energy sources as a foreign policy tool is seen as ineffective or outright dangerous (Carter & Nivola, 2009).

3.5.4   *Strategic Versus Market-Oriented Approach: Summary*

As indicated above, these approaches are naturally ideal models which, in pure form, define the endpoints of an axis that encompasses the dichotomy between state-guided versus market-guided energy policies. A real-world policy might be located at any point along the line connecting these two endpoints. It is noteworthy that these approaches are adopted in particular case-specific forms by different actors. The strategic approach is evident mainly in the state policies of producer states, the market-based approach in those of consumer states. The best example is given by contrasting the approach taken by the Russian Federation with that of the European Union. The former is a unitary actor with state-owned energy companies (often accused of operating on behalf of the home state and serving the state's needs); the latter actor is a quasi-state organization of sovereign states that try to enhance their position by creating a common area within which they share and address the risks of import dependence. Here, the market-based approach has been implemented. A quote from the former European Commissioner for Energy, Günther Oettinger, gives crystal-clear expression to the EU's dedication to the market-based approach: 'The Internal energy market today is our fundamental and most effective tool to provide security of supply. Only a fully functioning market is able to take adequate corrective measures in case of a disruption' (The European Files, 2011, p. 8).

The chief differences between the two approaches are outlined for clarity in Table 3.3.

Table 3.3   Strategic and market-based approaches: a comparison

	Strategic approach	*Market-based approach*
Theoretical basis	The realist tradition in IR, classic geopolitics	The liberal tradition in IR, neoclassical, and neo-institutional economics
General approach to energy policy in international relations	The need for independence from external supplies of energy	Energy independence is impossible, attempts to achieve it disrupt interstate relations
Management of energy resources	Scarcity of resources, which leads to resource nationalism	Market ensures efficient allocation
Role of energy policy in international relations	Utilized to influence international relations	Politicization of energy affairs leads to poor allocation and a less effective system
Main focus of energy policy	Emphasis on securing an adequate, secure supply, especially for oil and natural gas	Comprehensive view, looking at all resources, and looking at the functioning of markets, infrastructure, and influence
Nature of relations and distribution of resources	Zero-sum game	Non-zero-sum game
Patterns of cooperation in the international environment	International relations founded on bilateral relations; such a style is more predictable and is easier to influence	Cooperation with international organizations; multilateral relations are seen as less susceptible to influence and thus more stable
Positioning of actors in the international system	States as the main and only relevant actors	Multiple influential actors (including firms, international organizations, interest groups)
Role of the market	High risk of market failure, substantial role for the state	Supplies allocated effectively without state interference
Positioning of energy resources	Subject to the strategic interests of the state. They require special attention, mainly due to scarcity	Common market commodity
Future development	Possible conflict over energy resources and transit infrastructure	Scarcity of resources is best solved by cooperation among participating actors in the system
Optimal solutions	Independence or expansion	Interdependence by market means

Source: Adelman, 1973, p. 73; Burchill et al., 2005, pp. 30–54; Carter & Nivola, 2009; Chester, 2009, pp. 889–892; Ciuta, 2010, p. 128; Dannreuther, 2003, pp. 200–201; Fattouh, 2013; Klare, 2005, 2009a, 2009b; Leverett, 2009, pp. 213–227; Moran, 2009, pp. 19–23; Nordhaus, 2009; Tunsjø, 2010, pp. 25–45; Waltz, 1979, pp. 79–101. Compilation: the authors

3 RESEARCH DESIGN   41

3.6   Ideal Type Model for Assessing the Oil Sector

From the theoretical basis and differences between strategic and market-oriented approaches described above, the authors derived a set of features and indicators of strategic behaviour, adapted to the needs of this research, which will be used to assess the oil sector in the countries under investigation. These indicators were sought in data gathered from openly available sources and during semi-structured interviews with selected analysts and officials in the countries (see the above subsection on the research process). The authors are aware that no theoretical model is likely to appear in pure form in the real world. The model thus describes an ideal state or model of standardized behaviour. The degree of similarity, based on an assessment of how many indicators were pertinent to a particular case and which indicators those were, defined the case. In any case, the indicators occur within a complex context, and must be perceived accordingly. No indicator's presence by itself is enough to prove that energy relations are being misused for political purposes. It is rather the complex picture, the overall number of indicators present in the case, and context that provide a basis for outcomes (hence the use of intrinsic case studies, see above). In essence, individual indicators merely illuminate the developmental path in individual cases. Nevertheless, the model provided comparable results for case studies that help to highlight the differences among the individual cases (Table 3.4).

3.7   Explanation of Individual Indicators

Each indicator is subsumed under a broader feature class of the strategic behaviour involved. The explanation of individual features and indicators is as follows:

Feature: Energy as a tool of the state/the economy as a basis for state power. This feature is based on the assumption that economic output determines the power of the state. Energy is one of the key determinants of the state's economic output, as understood in the realist tradition of thought.

(a) Active support by Russian state representatives for energy enterprises and their activities abroad: The indicator is positive if Russian state representatives act or speak in favour of the interests of companies of Russian origin.

Table 3.4   Features and indicators of the ideal type model of strategic behaviour

Feature	Indicator
Energy as a tool of the state—the economy as a basis for state power	Active support by Russian state representatives for energy enterprises and their activities abroad
	As a foreign supplier, Russia rewards certain behaviours and links energy deals to the client state's foreign policy orientation
	Abuse of infrastructure (e.g. pipelines) and differential pricing to exert pressure on the client state
Energy resources perceived as strategically important and deserving special treatment	Efforts to take control of the energy resources, transit routes, and distribution networks of the client state
	Disruption (by various means) of alternative supply routes/sources of supply
Zero-sum game (against cooperation)	Efforts to gain a dominant market position in the client country
	Efforts to eliminate competitive suppliers
	Acting against liberalization
Relying on bilateral relations/ agreements	Diminishing the importance and influence of multilateral regimes such as the EU
Undesirable dependence (while increasing the dependence of others)	Attempts to control the entire supply chain (regardless of commercial rationale)
Emphasis on strategic issues (over economic logic)	Economically irrational steps taken to maintain a particular position in the client state's market

Source: The authors

(b) As a foreign supplier, Russia rewards certain behaviours and links energy deals to the client state's foreign policy orientation: The indicator is positive if conditions on supplies, energy prices, or related deals correlate with the state of relations between the state and Russia.

(c) Abuse of infrastructure (e.g. pipelines) and differential pricing to exert pressure on the client state: The indicator is positive if the supplier misuses its position within the supply infrastructure (e.g. charging high transit fees for competing suppliers, imposing trade barriers) to exert financial or supply pressure (e.g. supply curtailments, irregularities, hindering access of alternative suppliers) on the consumer.

Feature: Energy resources perceived as strategically important, deserving of special treatment. This feature relates mainly to non-renewable energy sources as traditional sources for the economy (mostly fossil fuels). In this sense, energy resources (as well as essentially the entire energy sector) are perceived as strategically important and deserving of special treatment (i.e. not to be entrusted to an uncontrolled market—hence the preference for state control and bilateral deals, see below). This feature assumes energy commodities are vital to state functioning and, given their finite nature, require special treatment. Such treatment is based on the realist understanding of a state's role (with the state as the key actor in the system). Therefore, if energy sources are vital to a state's survival, the state itself should take care of their distribution and make sure no other actor in the system (i.e. a competing state) acquires additional sources at the expense of the state in question.

(a) Efforts to take control of the energy resources, transit routes, and distribution networks of the client state: In line with the perception of the energy sector as strategically important, the indicator is positive if the supplier controls or aims to control key subareas of the client's energy sector: extraction, transport, and distribution.

(b) Disruption (by various means) of alternative supply routes/sources of supply: The indicator is positive if the supplier or dominant company aims at fending off potential competitors or sources of supply (suppliers as well as actual sources).

Feature: Zero-sum game. This feature relates to one of the main assumptions of the realist tradition and may be summarized as 'one person's gain is another person's loss'. In the energy sector, this relates mainly to the finite nature of energy resources (mainly fossil fuels): if one state acquires more resources, the remaining states get less, limiting their power.

(a) Efforts to gain a dominant market position in the client country: The indicator is positive if a company (in this case, one of Russian origin, often with the government's backing) clearly tries to dominate the market as a key goal.

(b) Efforts to eliminate competitive suppliers: Here, the indicator is positive if the goal of the company is not only to be the main player on the market but also to eliminate competitors.

(c) Acting against liberalization: This indicator is positive if the company opposes liberalizing the commercial environment in the country. This often happens with the backing of the country's home government. The motivation is that liberalization usually means tougher competition and erosion of the dominant position of the current incumbents.

Feature: Relying on bilateral relations/agreements. This feature relates to an assumption within the realist tradition that bilateral relations should be favoured over multilateral regimes, since they are easier influenced and steered to meet the dominant player's preferences.

(a) Diminishing the importance and influence of multilateral regimes such as the EU: The indicator is positive if opposition (by a company of Russian origin by itself or with the backing of the homeland government or its representatives) to liberalizing regimes is evident.

Feature: Undesirable dependence (while increasing the dependence of others)—The feature relates to an assumption of the realist tradition that a state should be able to survive by itself and not be dependent. At the same time, since the goal is actually to dominate other states (using the logic of the zero-sum game), making other actors dependent increases the state's power. In this sense, energy sources are tools for making this happen.

(a) Attempts to control the entire supply chain (regardless of commercial rationale): The indicator is positive if the company (often with the homeland government's backing) aims to assert control over the supply chain regardless of profit. Here, it is important to pay attention to context—some investments may not make sense in the short to mid-term but still be beneficial (financially and politically) in the long run.

Feature: Emphasis on strategic issues (over economic logic). This relates to the main interest of the realist tradition in the arena of international relations—the survival of the state and the struggle for power and dominance over others. In this sense, steps taken in acquiring certain goals need not be economically viable to be desirable.

(a) Economically irrational steps taken to maintain a particular position in the client state's market: The indicator is positive if the company (often with homeland government backing) holds a position regardless of its economic viability. Here, as with the previous indicator, the broader context must be taken into account, including potential future developments and eventual benefits.

3.8   Note on the Russian Oil Companies

Unlike in the natural gas sector, where the vast majority of activity is unified under OAO Gazprom, there are a number of Russian state-controlled companies operating in the oil sector of the Balkan countries, including PJSC Gazprom Neft (ОАО Газпром нефть), PAO NK Rosneft (ПАО Роснефть), PAO Transneft (ОАО Транснефть), and OAO Zarubezhneft (ОАО Зарубежнефть). This study follows the operational record of these companies, all the local subsidiaries these companies have created, and all the local companies these subjects are shareholders to. Their structures of shareholders are in Tables 3.5, 3.6, 3.7, and 3.8.

The study also follows the operational record of PAO Lukoil (ПАО Лукоил), a Russian company that is privately owned, as seen in Table 3.9.

The reason for including the company in the research is that its record of operations has raised serious suspicions of ties to the Russian establishment. The company's current major shareholder, Vagit Alekperov, was involved in its creation in the early 1990s. He was appointed Deputy Minister of the Oil and Gas Industry in the Soviet Union in 1990, and it was under his direction that Russia's first vertically organized oil company emerged. Out of the three Siberian companies Langepasneftegaz, Uralneftegaz, and Kogalymneftegaz, Alekperov created the LangepasUralKogalymneft state oil company in 1991 under the ministry. In 1993, its legal form was changed to that of an open joint-stock company, two refineries in Volgograd and Perm were added to its portfolio, and the name Lukoil was derived from the first letters of the three merged companies. Vagit

Table 3.5   OAO Gazprom Neft (ОАО Газпром нефть) shareholders

OAO Gazprom	95.68%
Minority shareholders (individuals and corporations)	4.32%

Source: OAO Gazprom Neft, n.d.

Table 3.6   OAO NK Rosneft (OAO HK Роснефтъ) shareholders as of 1 August 2016

Rosneftegaz AO[a]	69.5%
BP Russian Investments Limited	19.75%
National Settlement Depository	10.36%
Other legal entities with stakes lower than 5%	0.01%
Russian Federation (through the Federal Agency for State Property Management)	<0.01%
Individuals	0.38%
Unknown	<0.01%

Source: OAO NK Rosneft, n.d.

Note: Rosneft management has no information about shareholders whose equity stakes exceed 5% outside of those listed above

[a]Rosneftegaz AO is wholly owned by the Russian government. In December 2016, 19.5% of Rosneftegaz AO's shares were sold off. The new holder of these shares is QHG Shares Pte. Ltd. (a joint venture of Glencore plc and Qatar Investment Authority, the Qatar state-owned fund). Rosneftegaz AO kept a controlling share in the company by retaining a 50.00000001% equity stake

Table 3.7   OAO AK Transneft (OAO AK Транснефтъ) shareholders

Russian government	78.1%
Domestic and international investors[a]	21.9%

Source: OAO AK Transneft, n.d.

[a]The 21.9% stake consists of preferred shares with no voting rights

Table 3.8   OAO Zarubezhneft (Зарубежнефтъ OAO) shareholders

Federal Agency for Administration of State Property (Rosimushchestvo), a unit of the Ministry of Economic Development	100%

Source: OAO Zarubezhneft, n.d.

Alekperov has made use of his leading position in the Soviet Ministry of the Oil and Gas Industry and firm support from Gazprom founder Viktor Chernomyrdin, whose wish was to see a giant oil company with a similar profile in Russia. During his tenure as Russia's prime minister in 1992–98, Chernomyrdin acted as a protective godfather and was possibly an early shareholder (Gorst, 2007, p. 7). Despite its privately owned status, it has always been acutely aware of federal priorities (Grace, 2005, p. 220, cit. as in Aalto, Dusseault, Kivinen, & Kennedy, 2012, p. 24).

Table 3.9   PAO Lukoil (ПАО ЛУКОЙЛ) shareholders

Company management	**34.6%**
Vagit Alekperov	22.96%
Leonid Fedun	9.78%
Sergei Kukura	0.39%
Ravil Maganov	0.38%
Lyubov Khoba	0.35%
Alexander Matytsyn	0.30%
Valery Subbotin	0.20%
Sergei Mikhailov	0.06%
Vladimir Nekrasov	0.04%
Ivan Maslyaev	0.03%
Anatoly Moskalenko	0.02%
Valery Grayfer	0.01%
Vadim Vorobyov	0.01%
Ivan Pictet	0.009%
Evgeny Khavkin	0.008%
Gennady Fedotov	0.007%
Sergei Malyukov	0.005%
Azat Shamsuarov	0.005%
Denis Rogachev	0.001%
Richard Matzke	0.0003%
Guglielmo Antonio Claudio Moscato	0.00006%
Lukoil Investments Cyprus Ltd.[a]	**16.2%**
Other	**49.2%**

Source: PAO Lukoil, n.d.; PJSC 'LUKOIL', 2016, pp. 104–108, 118

Note: Other than the persons listed above, company management is not aware of any shareholders holding more than 5% of the company's charter capital. There are 42,713 investors in PAO Lukoil
[a]A wholly owned subsidiary of PAO Lukoil

A close relationship to the Russian government has been admitted by Alekperov himself, who said 'it is impossible to separate the interests of the company from the interests of the state on whose territory it operates. We have the same interests. Whatever is good for Russia is good for our company' (Korobochkin, 2004).

With the breakup of OAO NK Yukos (OAO HK ЮKOC) over alleged unpaid taxes—in a case constructed, according to all the evidence, to rid the establishment of Mikhail Khodorkovsky—a giant corporation responsible for 20% of Russia's oil production was liquidated. Alekperov accordingly lowered his profile and began to cultivate the image of a 'loyal' oilman, taking every opportunity to demonstrate his loyalty to Vladimir

Putin (Gustafson, 2012, pp. 319–358). Once Putin came to power, private oil companies had to tow the government line or risk extinction, as the case of OAO NK Yukos had demonstrated (Gorst, 2007, p. 11). Thus, all private oil companies in Russia fostered close ties with the state, including PAO Lukoil and OAO Surgutneftegaz (OAO Сургутнефтегаз) (Mankoff, 2009, p. 9, cited in Smith, 2012, pp. 39–62).

The major motivation for including PAO Lukoil in the study, then, was the information available on the company, along with the allegations and accusations surrounding its operations. An unbiased picture of the relationship between PAO Lukoil and the Russian government is difficult to obtain; there is no decisive evidence for whether it acts as a private company or in concert with the Russian government. What is clear, however, is that the company does not act against its government's interests. Corollary to this, there are 42,713 investors in PAO Lukoil whose names are not disclosed unless they hold more than 5% of the company. Some current board members have personal ties to the state, and some are former cabinet ministers (Valery Grayfer, Igor Ivanov) (Koďousková, Kuchyňková, Leshchenko, & Jirušek, 2014, pp. 173–175). However, the precise nature of the personal ties between PAO Lukoil and representatives of the Russian establishment remains unclear and doubtful.

The operations of PAO Lukoil are therefore scrutinized in the study together with those of the state-owned enterprises. The reader should approach the study with knowledge of the companies' shareholder structures and keep in mind that the relationship between PAO Lukoil and the Russian government must be approached with caution.

Sources

Aalto, P., Dusseault, D., Kivinen, M., & Kennedy, M. D. (2012). Chapter 2: How Are Russian Energy Policies Formulated? Linking the Actors and Structures of Energy Policy. In P. Aalto (Ed.), *Russia's Energy Policies. National, Interregional and Global Levels* (pp. 20–42). Cheltenham: Edward Elgar Publishing. https://doi.org/10.1111/1478-9302.12016_41

Adelman, M. A. (1973). *The World Petroleum Market*. Baltimore: The John Hopkins University Press.

Brinkmann, S. (2013). *Qualitative Interviewing: Understanding Qualitative Research*. Oxford: Oxford University Press.

Burchill, S., et al. (2005). *Theories of International Relations*. Houndmills: Palgrave Macmillan.

Carter, E., & Nivola, P. (2009). Making Sense of "Energy Independence". In J. P. Elkind (Ed.), *Energy Security: Economics, Politics, Strategies and Implications* (pp. 105–116). Washington, DC: Brookings Institution Press.

Chester, L. (2009). Conceptualising Energy Security and Making Explicit Its Polysemic Nature. *Energy Policy, XXXVIII*, 887–895.

Ciuta, F. (2010). Conceptual Notes on Energy Security: Total or Banal Security? *Security Dialogue, XLI*(2), 123–144.

Creswell, J. W. (2009). *Research Design: Qualitative, Quantitative and Mixed Methods Approaches.* Lincoln: Sage.

Dannreuther, R. (2003). Asian Security and China's Energy Needs. *International Relations of the Asia-Pacific, 3*(2), 197–219.

Fattouh, B. (2013, February 15). *The Financialization of Oil Markets: Potential Impacts and Evidence.* Retrieved from https://www.hhs.se/contentassets/dec6de1bee5b433093b74bb766a6b2ac/financialization-of-crude-oil-markets-sweden.pdf

Gerring, J. (2007). *Case Study Research: Principles and Practices.* New York: Cambridge University Press.

Gilpin, R. (2001). *Global Political Economy: Understanding the International Economic Order.* Princeton: Princeton University Press.

Gorst, I. (2007). *Lukoil: Russia's Largest Oil Company.* James A. Baker III Institute for Public Policy Research Paper. Houston: Rice University. Retrieved from http://www.bakerinstitute.org/files/3902/

Gustafson, T. (2012). *Wheel of Fortune. The Battle for Oil and Power in Russia.* Cambridge and London: The Belknap Press of Harvard University Press.

Holsti, O. R. (1969). *Content Analysis for the Social Sciences and Humanities.* Reading, MA: Addison-Wesley Publishing Company.

Jackson, R., & Sorensen, G. (2007). *Introduction to International Relations: Theories and Approaches.* Oxford: Oxford University Press.

Jirušek, M. (2017). *Politicization in the Natural Gas Sector in South-Eastern Europe: Thing of the Past or Vivid Present?* (1st ed.). Brno: Masaryk University.

Jirušek, M., & Kuchyňková, P. (2018). The Conduct of Gazprom in Central and Eastern Europe: A Tool of the Kremlin, or Just an Adaptable Player? *East European Politics and Societies, 32*, 818–844. https://doi.org/10.1177/0888325417745128

Jirušek, M., Vlček, T., & Henderson, J. (2017). Russia's Energy Relations in Southeastern Europe: An Analysis of Motives in Bulgaria and Greece. *Post-Soviet Affairs, 33*(5), 335–355. https://doi.org/10.1080/1060586X.2017.1341256

Jirušek, M., Vlček, T., Koďousková, H., Robinson, R. W., Leshchenko, A., Černoch, F., et al. (2015). *Energy Security in Central and Eastern Europe and the Operations of Russian State-Owned Energy Enterprises.* Brno: Masaryk

University. ISBN 978-80-210-8048-5. Retrieved from https://munispace. muni.cz/library/catalog/book/790

King, G., Verba, S., & Keohane, R. O. (1994). *Designing Social Inquiry: Scientific Inference in Qualitative Research*. Princeton: Princeton University Press.

Klare, M. T. (2014). *Twenty-first Century Energy Wars: How Oil and Gas Are Fuelling Global Conflicts*. Retrieved from http://www.energypost.eu/twenty-first-century-energy-wars-oil-gas-fuelling-global-conflicts/

Klare, M. T. (2005). *Blood and Oil: The Dangers and Consequences of America's Growing Dependency on Imported Petroleum*. New York: Holt Paperbacks.

Klare, M. T. (2009a). There Will Be Blood: Political Violence, Regional Warfare and the Risk of Great-Power Conflict over Contested Energy Sources. In A. Korin & G. Luft (Eds.), *Energy Security Challenges for the 21st Century: A Reference Handbook* (pp. 44–61). Santa Barbara: Praeger Security International.

Klare, M. T. (2009b). Petroleum Anxiety and the Militarization of Energy Security. In D. Moran & J. A. Russel (Eds.), *Energy Security and Global Politics: The Militarization of Resource Management* (pp. 39–61). New York: Routledge.

Koďousková, H., Kuchyňková, P., Leshchenko, A., & Jirušek, M. (2014). *Energetická bezpečnost asijských zemí a Ruské federace (Energy Security of the Asian Countries and the Russian Federation)*. Brno: Masaryk University.

Kořan, M. (2008). Jednopřípadová studie. In P. Drulák (Ed.), *Jak zkoumat politiku* (pp. 29–61). Praha: Portál.

Korobochkin, M. (2004, August 5). Триумф <тихого магната>. *Люди*. Retrieved from http://www.peoples.ru/undertake/hard/alekperov/

Krippendorf, K. (2013). *Content Analysis: An Introduction to Its Methodology*. Sage Publications.

Leverett, F. (2009). Resource Mercantilism and the Militarization of Resource Management. In D. Moran & J. A. Russel (Eds.), *Energy Security and Global Politics: The Militarization of Resource Management* (pp. 211–242). New York: Routledge.

Levy, J. S. (2008). Case Studies: Types, Designs, and Logics of Inference. *Conflict Management and Peace Science, 25*, 1–18.

Luft, G., & Korin, A. (2009). Realism and Idealism in the Energy Security Debate. In G. Luft & A. Korin (Eds.), *Energy Security Challenges for the 21st Century: A Reference Handbook* (pp. 335–348). Santa Barbara: Praeger Security International.

Moran, D. (2009). The Battlefield and the Marketplace: Two Cautionary Tales. In D. Moran & J. A. Russel (Eds.), *Energy Security and Global Politics: The Militarization of Resource Management* (pp. 19–33). New York: Routledge.

Nordhaus, W. (2009, June 17). *The Economics of an Integrated World Oil Market*. Keynote Address, International Energy Workshop, Venice, Italy. Retrieved from http://www.econ.yale.edu/~nordhaus/homepage/documents/iew_052909.pdf

OAO AK Transneft. (n.d.). Retrieved from http://transneft.ru/

OAO Gazprom Neft. (n.d.). Retrieved from http://www.gazprom-neft.com/

OAO NK Rosneft. (n.d.). Retrieved from https://www.rosneft.com/

OAO Zarubezhneft. (n.d.). Retrieved from http://www.zarubezhneft.ru/

Odell, J. S. (2004). Case Study Methods in International Political Economy. In D. F. Sprinz & Y. Wohlinsky-Nahmias (Eds.), *Models, Numbers & Cases: Methods for Studying International Relations* (pp. 56–80). Ann Arbor: University of Michigan Press.

PAO Lukoil. (n.d.). Retrieved from http://www.lukoil.com/

PJSC "LUKOIL". (2016). *PJSC "LUKOIL" Annual Report 2015.* Retrieved from http://www.lukoil.com/materials/doc/Annual_Report_2015/LUKOIL_AR_2015_ENG.pdf

Smith, H. (2012). Domestic Influences on Russian Foreign Policy: Status, Interests and Ressentiment. In M. R. Freire & R. E. Kanet (Eds.), *Russia and Its Near Neighbours* (pp. 39–62). Basingstoke: Palgrave Macmillan.

Stake, R. E. (2006). *Multiple Case Study Analysis.* New York: The Guilford Press.

The European Files. (2011, May–June). *Security of Supply in Europe: Continuous Adaptation.* Retrieved from http://europeanfiles.eu/wp-content/uploads/issues/2011may-june.pdf

Tunsjø, Ø. (2010, March). Hedging Against Oil Dependency: New Perspectives on China's Energy Security Policy. *International Relations, XXIV,* 25–45.

Waltz, K. (1979). *Theory of International Politics.* Boston: McGraw-Hill.

Yergin, D. (2005, July 31). It's Not the End of the Oil Age. *The Washington Post.* Retrieved from http://www.washingtonpost.com/wp-dyn/content/article/2005/07/29/AR2005072901672.html

Yergin, D. (2006, December 17). A Great Bubbling. *Newsweek.* Retrieved from http://www.newsweek.com/great-bubbling-105841

CHAPTER 4

Albania

4.1 Crude Oil Sector General Information

4.1.1 Introduction and Upstream

Albania is a small country in the Western Balkans with 2.9 million inhabitants. Neighbouring Montenegro, Kosovo, Former Yugoslav Republic of Macedonia (FYROM), and Greece, the country is separated on the West from Italy by the Adriatic Sea, with Croatia a short distance to the North. Albanian crude oil consumption is miniscule: only 255,000 tonnes in 2014 (Gjika & Toro, 2016). In 2013, oil and petroleum products constituted 58% of Albania's total primary energy supply, and a value of around 60% appears valid for the long term as well (Energy Community, n.d.; Leskoviku, 2003). Because it is able to produce its own domestic crude oil, and because there are problems in the refining sector, Albania does not import crude oil, but it does export it: crude oil was responsible for around 30% of total exports in 2013. Albanian crude oil is currently exported by small tankers to Spain, Italy, Greece, and Malta (Gjika & Toro, 2016; Popovici, Deliso, Michaletos, & Stavros, 2012). It does import refined petroleum products; currently nearly 100% of the oil products consumed come from abroad, approximately 1 million tonnes annually (mta; Interview 01). Though current oil extraction levels of around 1.3 mta[1] are

---

[1] Oil extraction: 739,000 tonnes in 2010, 1.028 million tonnes in 2012, 1.205 million tonnes in 2013, and 1.368 million tonnes in 2014 (Albpetrol Sh.A., n.d.).

© The Author(s) 2019
T. Vlček, M. Jirušek, *Russian Oil Enterprises in Europe*,
https://doi.org/10.1007/978-3-030-19839-8_4

enough to satisfy Albania's domestic needs, since the closure of the Ballsh refinery domestic oil product output has been limited, and the excess demand has been made up by imports. As a result, refined petroleum products now constitute 15% of Albania's total imports (Vlček & Jirušek, 2018, p. 99).

Crude oil extraction has a long history in Albania, with the first modern wells becoming operational in the 1930s. During the socialist leadership of Enver Halil Hoxha in Albania from 1941 to 1985, production attained values of up to 2.25 mta (mainly thanks to close economic cooperation with China), which covered domestic consumption needs and allowed for exports. This changed considerably in the late 1980s, when production dropped by half due to changes in diplomatic relations with China, eventually declining to 0.475 mta in 1994. Further production declines (bottoming out at 0.315 mta in 2000) followed, but Albania eventually managed to attract foreign investors to its oil production sector in the 2000s (Albpetrol Sh.A., n.d.; Popovici et al., 2012). The adoption of Act No. 7746 for Research, Development, Production, Handling, Transportation, Storage, and Sale of Crude Oil, Gas, and Tar Sands Within and Outside the Republic of Albania in 1993 was especially helpful. Currently, five foreign companies are engaged in crude oil production in Albania (see Table 4.1) (Vlček & Jirušek, 2018, p. 100).

The National Agency of Natural Resources (Agjencia Kombëtare e Burimeve Natyrore, AKBN), subordinate to the Ministry of Economy, Trade, and Industry, is the overarching body that oversees compliance with Act No. 7746 and administers the oilfields. The National Agency of Natural Resources negotiates and licenses Petroleum Sharing Contracts (PSCs) with foreign entities to develop and produce petroleum on behalf of the Albanian Ministry of Economy, Trade, and Industry.[2]

The only domestic oil production company is Albpetrol Sh.A., based in Patos, which is fully owned by the Ministry of Economy, Trade, and Industry. It also operates through Servcom Sh.A. (Service Company of Oil and Gas), which provides support services for the Albanian oil industry and explores and extracts crude oil and natural gas.

---

[2] Under Act No. 7746, the Albanian state, which owns all the oil and gas reserves in the country and is represented by the National Agency of Natural Resources, can enter into Production Sharing Contracts with state or private companies. These PSCs give exclusive rights to the state's partner to explore and produce oil and gas within a defined perimeter for 25 years (Albpetrol Sh.A., n.d.; Popovici et al., 2012).

Table 4.1 Companies active in the Albanian Crude Oil Extraction Sector as of 2016

Company	Ownership	Production	Oil production share (%)	Company's active oilfields
Bankers Petroleum Albania Ltd.	Bankers Petroleum Ltd. (100%)	1212.84	88.6	Patos-Marinza
TransAtlantic Petroleum Ltd.	Malone Mitchell (36%), Nokomis Capital, L.L.C. (10%) other shareholders	81.23	5.9	Ballsh-Hekal, Cakran-Mollaj, Gorisht-Kocul
Sherwood International Petroleum Ltd.	Bankers Petroleum Ltd. (100%)	0.19	<0.1	Kuçovë
Transoilgroup Sh.A.	Transoilgroup AG (100%)	24.62	1.8	Visokë
Phoenix Petroleum Sh.A.	Phoenix Petroleum Ltd. (100%)	5.07	0.4	Amonicë
Albpetrol Sh.A.	Ministry of Economy, Trade and Industry of the Republic of Albania (100%)	45.38	3.3	Ballsh, Patos, Kuçovë, Amonicë

Source: Albpetrol Sh.A., n.d.; Bankers Petroleum Ltd., n.d.; National Agency of Natural Resources, n.d.; Phoenix Petroleum Sh.A., n.d.; TransAtlantic Petroleum Ltd., n.d.; Transoilgroup AG, n.d.; Vlček & Jirušek, 2018, p. 101

Note: Figures in thousands of tonnes annually, data from 2014; until 2014, Stream Oil & Gas Ltd. was active in the Albanian Crude Oil Extraction Sector; the company now operates under TransAtlantic Petroleum Ltd., having acquired Stream Oil & Gas Ltd. in November 2014

Albpetrol Sh.A. was to be privatized to further increase oil production in the country by attracting a strategic partner with the resources and know-how to undertake oilfield rehabilitation, to enhance oil recovery, and to explore new areas (Popovici et al., 2012). The company was offered for privatization in an international tender in 2012; the highest bid of EUR 850 million came from Vetro Energy Ltd. controlled by Rezart Taçi (over Win Business Petroleum Group Ltd. from China, and Bankers Petroleum Ltd. from Canada). Thirteen countries expressed interest, including PAO Gazprom, whose offer of EUR 50 million was the lowest. Since the price of the physical assets of the company was several times greater than PAO Gazprom's offer, it was not accepted (Interview 01; Vlček & Jirušek, 2018, p. 102).

Vetro Energy Ltd., the winner of the privatization, failed to make good on its 20% guarantee and the deal was called off by Albanian officials ("*Albania Calls*", 2013; "*Albania Probes*", 2013; "*Albania Tycoon*", 2013; "*Vetro Says*", 2012). Word is that after the cancellation, Albania attempted to negotiate the privatization with Win Business Petroleum Group Ltd. (the second-highest bidder), but the deal was not closed.

Several countries have, however, concluded Albpetrol Sh.A. Petroleum Sharing Contracts since 2004. The first was Canada-based Bankers Petroleum Ltd.[3] for Patos-Marinza, the largest onshore oil field in Europe, situated in central Albania, and later, via the acquisition of Sherwood International Petroleum Ltd., the Kuçova fields. These fields are well developed and oil extraction is extensive. Current foreign explorers and prospectors include Petromanas Energy Inc.[4] (Canada), San Leon Energy PLC (Ireland), Emanuelle Adriatic Energy Ltd. (Israel), and Pennine Petroleum Corporation (Canada) (Vlček & Jirušek, 2018, p. 102).

Petromanas Energy Inc. holds three PSCs in six onshore blocks in Berati. Recently, it found partners for a joint venture to develop these blocks. Petromanas was originally in possession of 25% of shares and had operator status, while Royal Dutch Shell plc owned 75% of shares and had investor status (Albpetrol Sh.A., n.d.). Since February 2016, Royal Dutch Shell has been the 100% owner of Petromanas. San Leon Energy PLC is active in exploration of the large offshore oil and gas Durrësi Block (Beach Energy Limited based in Australia originally held a 25% stake in the Durrësi Block PSC but withdrew in 2011, making San Leon Energy PLC the 100% shareholder). Emanuelle Adriatic Energy Ltd. has one offshore PSC in the Adriatiku 2 and 3 locations in the Adriatic Sea. There is also Capricorn Albania Limited (owned by Scotland-based Cairn Energy PLC), which closed one PSC in 2007 for seven years. The evidence, however, shows the Joni Offshore Block has not been developed after the PSC's expiration.

The Albanian strategy appears to be to elevate the country's regional importance by enhancing extraction activity with the aim of exporting oil

---

[3] In March 2016, it was announced that Bankers Petroleum Ltd. would be sold to Alberta Ltd. and Charter Power Investment Limited, both subsidiaries of Geo-Jade Petroleum Corporation, one of the largest Chinese independent oil and gas exploration and production companies. The transaction price is C$ 545 million, that is, US $442 million ("*Bankers Petroleum Ltd. Enters*", 2016).

[4] Petromanas Energy Inc. sold all its assets to Royal Dutch Shell plc in February 2016 for US $45 million (Koleka & Weir, 2016).

to Italy, Spain, and Greece on a regular basis, and possibly to other countries as well. If the country does indeed export to Italy, that might mitigate the need for the AMBO pipeline and, now that Bulgaria has withdrawn from the Burgas-Alexandroupolis project, arouse interest from Greece to get oil from a neighbouring country[5] (Çukaj, 2016, p. 37; Popovici et al., 2012). Oil production is crucial to generating revenue from the oil industry for Albania. Its economy was boosted by US $400–500 million in 2012–2013 (Çukaj, 2016, p. 36); in 2014, revenues reached US $583 million, with Bankers Petroleum Albania Ltd. alone generating about 5.5% of the country's gross domestic product (Erebara, 2015). The Albanian economy and its regional power could substantially benefit from oil exploration and export income.

4.1.2   Midstream

When it comes to midstream, there is no international oil pipeline in Albania. This is mainly because the crude oil the country consumes was formerly extracted within its own borders and then refined in local refineries. Two crude oil pipelines from the Ballsh (70 km) and Fier (35 km) refineries to the Vlorë maritime oil terminal have been out of service because of their poor condition (Energy Community, 2011, p. 33). A pipeline from the Cërrik refinery through Kuçovë to the Vlorë maritime terminal was used in the past, but it is old and fell out of use when the Cërrik refinery was shut down in 1990 (Interview 03). Bankers Petroleum Albania Ltd. is currently constructing a 35 km export crude oil pipeline from the central Fier hub loading facility to Vlorë (construction is being handled by the Dutch firm A.Hak Pipelines & Facilities B.V.). The pipeline's role is to transport oil from the company's Patos-Marinza field to the export terminal more cheaply. It was to become operational by 2016 (though there is no information on a completion date), with a planned capacity of 3.5 mta (70,000 barrels per day) (Bankers Petroleum Ltd., 2013; Fawcett, 2012). Indications are that it is in fact the reconstruction of the old pipeline running from the Fier refinery to Vlorë (Vlček & Jirušek, 2018, p. 103).

---

[5] For more information about the AMBO and Burgas-Alexandroupolis projects, see the chapters on Bulgaria and Greece.

4.1.3   Downstream

The Albanian refinery sector is organized under Albanian Refining & Marketing of Oil Sh.A. (ARMO). Established in 1999 by the Albanian government as a state-owned corporation, it owns two refineries in Albania: Ballsh with 1.0 mta design capacity and Fier with 0.5 mta design capacity. ARMO is also in possession of the Vlorë maritime oil terminal and 11 depots in the country. ARMO is owned by AMRA Oil (85%) and the Government of Albania (15%).

The third Cërrik refinery was the oldest in the country but was shut down in 1990 for environmental reasons and later dismantled. The Ministry of Economy, Trade, and Industry tried to privatize the facility, and former Minister Ilir Rexhep Meta strongly supported its sale in 2011 (*"Cerrik Oil"*, 2011). Kürüm International Sh.A. (an Albanian subsidiary of Turkey's Kürüm Holding A.Ş.) won the procurement procedure and was willing to rebuild the refinery (which had been completely dismantled; Interview 06) and restart operations, but in 2013 the Tirana court suspended the deal after a rival bidder alleged irregularities. Kürüm International Sh.A. operates a steel plant (Kombinati Metalurgjik) in Elbasan, Albania, and under a public procurement procedure acquired four operational hydroelectric power plants (Ulëza, Shkopet, Bistricë I, and Bistricë II, together known as HEPPs) with a total installed capacity of 76.5 MWe in 2013. Kürüm International Sh.A.'s interest in the Cërrik refinery is part of its broader entry into the Albanian industrial market. But the company's plans for expansion are now in jeopardy—it was forced to file for bankruptcy in March 2016 because of intense competition from cheap steel, mainly from China (Interview 05; Kürüm Holding A.Ş., n.d.; *"Tirana Court"*, 2016; Vlček & Jirušek, 2018, pp. 105–106) (Table 4.2).

The Fier refinery has been operating for quite some time at around 20% of its designed refining capacity (*"Energy Profile of Albania"*, 2009), as was the Ballsh refinery until its eventual closure in 2014. A wide variety of products were produced at Ballsh (gasoline, gas oil, jet fuel, diesel oil, liquid petroleum gas [LPG], and heavy condensates). But the ageing of the technology eventually led to limits on the production of gas oil, heavy condensates, and diesel oil (Interview 03). Between 2014 and 2016, the only refinery still in operation in Albania was the Fier refinery, which functioned irregularly at around 20% of its design capacity, as noted, but remained profitable, producing heavy products such as fuel oil and bitumen. The focus on heavy condensates is chiefly dictated by

Table 4.2 Capacity of Albanian refineries

Refinery	Owner	Refining capacity	Type	Year established
Ballsh	Albanian Refining & Marketing of Oil Sh.A. (ARMO)	1.0 mta	Atmospheric distillation, delayed coking, hydrodesulphurization, catalytic reforming	1978 (closed in 2014, reopened in 2016)
Fier	Albanian Refining & Marketing of Oil Sh.A. (ARMO)	0.5 mta	Atmospheric Distillation, vacuum distillation, diluent production	1968
Cërrik	Ministry of Economy, Trade, and Industry of the Republic of Albania	0.55 mta	–	1956 (closed in 1990)

Source: Vlček & Jirušek, 2018, p. 106

the efforts to maintain competitiveness: the production of lighter products was given up to allow profitable operations with the ageing equipment base (Interview 02).

The technology used at both refineries has served as an economic limiting factor in itself. The Fier refinery was built in 1968 with Soviet technology; the Ballsh refinery in 1978 using already outdated Chinese technology from the 1960s. It was designed to refine heavy viscous crudes with high bitumen and sulphur content from domestic production only (International Energy Agency, 2008, p. 144). This prevented the import of other crude oil types and is the reason Albania still does not import crude oil (Vlček & Jirušek, 2018, p. 107).

Albanian Refining & Marketing of Oil Sh.A. has allegedly been used to finance political parties and campaigns, and this is a key reason for its poor economic performance and huge debts today (Interview 05; see later). These debts are what has led to the refineries' closure and limited operations. For more than a year starting in the summer of 2015, employees of the company received no salary (Interview 01), and ARMO was in serious trouble (Vlček & Jirušek, 2018, p. 107).

National Agency of Natural Resources data show that Albania produced 1.368 mta of crude oil in 2014, of which 255,000 tonnes was distributed in the internal market and 1.057 mta was exported (Gjika & Toro, 2016).

As it stands, Albania imports 100% of its gasoline, LPG, and jet fuel, and 60–70% of its diesel. Albanian Refining & Marketing of Oil Sh.A. is responsible for wholesale marketing of indigenous resources. Oil is predominantly consumed in the transport sector. When it comes to retail, the following companies have the strongest presence on the Albanian market: Alpet (a subsidiary of Turkey's Altınbaş Holding), Kastrati Sh.A., Kaspetrol Sh.A., Europetrol Durrës Albania Sh.A., Porto Romano Oil Sh.A., Bolv Oil Sh.A., Genklaudis Sh.A., Everest Oil Sh.A., Taçi Oil Sh.A., Armo Sh.A., and many others such as GEGA Oil, IP Albania, Gulf Albania, and so on (Vlček & Jirušek, 2018, p. 108).

The top companies in terms of retail market share are ALPET (around 100 petrol stations), Kastrati Sh.A. (over 70 petrol stations), and Everest Oil Sh.A. and Europetrol Durrës Albania Sh.A. (over 40 petrol stations) (Competition Commission of the Republic of Albania, 2015; Europetrol Durrës Albania Sh.A., n.d.; "*Karburantet, pse nuk*", 2015; Kastrati Sh.A., n.d.).

Oil products are also imported, mainly from Greece and in lesser quantities from Italy, Russia, and other countries. Imports chiefly flow through the ports of Vlorë, Durrës, Shëngjin, and Sarandë ("*Energy Profile of Albania*", 2009), all of which are capable of handling liquid bulks.

4.2   Recent Market Developments and Russian Activities

Albanian Refining & Marketing of Oil Sh.A. (ARMO) was originally wholly state-owned but underwent a privatization process, and ownership has changed hands twice since 2008. First, 85% of company shares were sold to the US-Swiss group AMRA Oil Consortium in 2008 for US $128.7 million on the basis of a public tender. Ownership was thus AMRA Oil (85%) and the Government of Albania (15%). AMRA Oil is owned by Refinery Associates of Texas, Mercuria Energy Group Ltd., and Anika Enterprises SA[6] (Anika's share has been estimated at 80%) through the company Anika Mercuria Refinery Associated Oil. Rezart Taçi, an Albanian businessman and owner of Anika Enterprises SA through Taçi Oil International Trading & Supply Company Sh.a., financed the purchase via the Azerbaijani state bank at EUR 75 million but the bank sued Taçi for failure to pay and demanded the seizure of ARMO's assets. A few days

[6] Named after Rezart Taçi's sister, Anika Taçi.

later, the demand was withdrawn ("*ARMO Changes*", 2013; "*ARMO Oil Refiner*", 2016; "*Debts Force*", 2013; Vlček & Jirušek, 2018, p. 105).

In 2013, Azerbaijani-based Heaney Assets Corporation took over Anika Enterprises SA's 80% shares in AMRA Oil for EUR 35 million and also paid almost EUR 15 million for minority stakes in several other companies controlled by the tycoon, including Serbian Euro Petrol d.o.o., Albanian Taçi Oil International Trading & Supply Company Sh.a. and Kuid Sh.p.k. ("*ARMO Oil Refiner*", 2016; "*ARMO Refiner*", 2014; "*Debts Force*", 2013; "*Will the New Broom*", 2013). This takeover was tied to Azerbaijan's claim that it was due EUR 75 million, with Heaney Assets Corporation representing Azerbaijan's interests.

The Albanian refining sector is strongly underfinanced to the point that it faces bankruptcy. The country's refineries are, as noted earlier, organized under ARMO Sh.A., owned by AMRA Oil (85%) and the Government of Albania (15%). To the company's recent history of privatization may be added the fact that the Government of Albania has attempted to sell its remaining 15% share. An auction was held in February 2016 with a price tag of around EUR 20 million, but Enforcement Private Service Albania Ltd., which held the auction, failed to attract any investors. Under Albanian law, the auction must now proceed to a second round at 50% of the original price, that is, EUR 10 million. The most probable reason for the lack of bidders was ARMO's huge debts. They amount to around US $267 million, of which US $209 million is owed to several foreign and domestic banks, and the rest to the Albanian tax office. The largest creditor among foreign entities is Raiffeisen Bank (in fact, the auction was meant to settle up the debt with Raiffeisen), and the largest local creditor is Albpetrol Sh.A. (Interview 02). The new owner, Heaney Assets Corporation, has apparently committed to pay off the company's debts, but has yet to take any concrete steps in doing so. The company's current situation is influenced to a substantial degree by the illegal business operations (credit fraud, tax fraud, etc.) of Taçi, its first owner after the 2008 privatization ("*Prapaskenat e kolapsit*", 2016; Vlček & Jirušek, 2018, p. 109).

The combined effect of technological limitations and the age of technology, along with the company's economic condition after the illegal syphoning of funds and its inability to attract investors, is pushing the refineries inexorably towards bankruptcy and closure. Omitting the obvious difficulties posed for employment policy, this would force Albania to import all the oil products it consumes from abroad and, one imagines, induce economic hardship in the process. But a large share of Albania's refined petroleum products have in fact already been coming from beyond its borders. The capacity of the maritime ports of Vlorë, Durrës, Shëngjin, and Sarandë is sufficient, and the country will likely entirely give up its domestic refining sector (Vlček & Jirušek, 2018, p. 110).

Despite this, in September 2016 the biggest oil producer in Albania, Bankers Petroleum Albania Ltd., suddenly closed a contract with Ionian Refining and Trading Co. (IRTC) Sh.A to supply crude oil to ARMO, when IRTC became the new owner of ARMO. This company, an offshore firm based in the British Virgin Islands, has an undisclosed shareholder structure administered by Besnik Sulaj, one of Albania's wealthiest citizens ("*LISTA E PLOTE*", 2015; "*Offshore Company*", 2016). Since ARMO's debts are estimated at US $600 million, the situation has not yet firmed up. ARMO is in the process of a complicated restructuring which has given rise to social unrest, with redundant workers being laid off and the company refusing to pay salaries that are due since ARMO's bankruptcy under the previous owner (Interview 08). Given the ageing technology, the quality of the fuels produced at ARMO's installations is also the lowest in the whole Balkan Peninsula (Vlček & Jirušek, 2018, p. 110).

4.3   Research Indicator Assessment

With the exception of very small quantities of oil products imported on a market basis from Russia, Russian Federation companies have no presence on the Albanian market. Because of this, the research indicator assessment is not applicable to this case study.

Sources

Albania Calls Off Vetro Offer for Albpetrol Oil Firm. (2013, February 8). *Reuters. com*. Retrieved from http://www.reuters.com/article/albania-albpetrol-sale-idUSL5N0B8DJ320130208

Albania Probes Failed Oil Company Privatization. (2013, January 6). *BalkanInsight. com*. Retrieved from http://www.balkaninsight.com/en/article/albania-probes-failed-oil-company-privatization

Albania Tycoon Suspected of Money Laundering. (2013, December 24). *BalkanInsight.com*. Retrieved from http://www.balkaninsight.com/en/article/albania-oil-tycoon-suspected-of-money-laundering

Albpetrol Sh.A. (n.d.). Retrieved from http://www.albpetrol.al/

ARMO Changes Owners. (2013, August 8). *Top-Channel.tv.* Retrieved from http://top-channel.tv/english/artikull.php?id=9724

ARMO Oil Refiner Key Assets on Sale for €20mln After Loan Default. (2016, February 12). *Tirana Times.* Retrieved from http://www.tiranatimes.com/?p=126146

ARMO Refiner Fined €35,000. (2014, December 29). *Tirana Times.* Retrieved from http://www.tiranatimes.com/?p=430

Bankers Petroleum Ltd. (2013, March 14). *March 2013 Corporate Presentation.* Calgary: Bankers Petroleum Ltd.

Bankers Petroleum Ltd. (n.d.). Retrieved from http://www.bankerspetroleum.com/

Bankers Petroleum Ltd. Enters into Definitive Agreement to Be Acquired by an Affiliate of Geo-Jade Petroleum Corporation. (2016, March 20). *PR Newswire.* Retrieved from http://www.prnewswire.com/news-releases/bankers-petroleum-ltd-enters-into-definitive-agreement-to-be-acquired-by-an-affiliate-of-geo-jade-petroleum-corporation-572824801.html

Cërrik Oil Refinery Offered for Privatization. (2011, August 5). *Top-Channel.tv.* Retrieved from http://top-channel.tv/english/artikull.php?id=2080

Competition Commission of the Republic of Albania. (2015, February 12). *Decision No. 345 of 12 February 2015 On Closing the In-depth Investigation into the Market of Production, Importing and Wholesale Selling of Fuels Against Undertakings Kastrati SH.A, Kaspetrol SH.A, Europetrol Durrës Albania SH.A, Porto Romano Oil SH.A, Bolv Oil SH.A, Genklaudis SH.A, Everest Oil SH.A, Taci Oil SH.A. and Armo SH.A. and Giving Recommendations on the Good Functioning of This Market.* Retrieved from http://www.caa.gov.al/uploads/decisions/Decision_345.pdf

Çukaj, I. (2016). Economic Security as Priority, Energy Security, Advantage of Western Balkans and Albania. *European Journal of Business, Economics and Accountancy, 4*(1), 31–38. Birmingham: Progressive Academic Publishing.

Debts Force Albanian Tycoon to Sell Assets. (2013, August 28). *BalkanInsight.com.* Retrieved from http://www.balkaninsight.com/en/article/debts-lead-to-downfall-of-albania-tycoon

Energy Community. (2011). *Emergency Oil Stocks in the Energy Community Level.* Final Report by Energy Community, Petroleum Development Consultants Ltd., and Energetski Institut Hrvoje Požar d.o.o. Retrieved from https://www.energy-community.org/portal/page/portal/ENC_HOME/DOCS/2516177/0633975AB2907B9CE053C92FA8C06338.PDF

Energy Community. (n.d.). Retrieved from https://www.energy-community.org/

Energy Profile of Albania, Energy Policy, Major Market Players, Energy Sources. (2009). *South-East European Industrial Market,* 5, (November–December), pp. 6–11. Retrieved from http://see-industry.com/img/industrial/240111011358SEEIM_2009_5.pdf

Erebara, G. (2015, March 13). Albanian Oil Production Registers Peak in 2014. *BalkanInsight.com*. Retrieved from http://www.balkaninsight.com/en/article/bankers-petroleum-albania-achieved-highest-production-ever-in-2014

Europetrol Durrës Albania Sh.A. (n.d.). Retrieved from http://www.europetrol-albania.com/

Fawcett, M. (2012, September 4). Against All Odds, Bankers Petroleum Is Trying to Produce Oil in Albania. *Alberta Venture*. Retrieved from http://albertaventure.com/2012/09/bankers-petroleum-is-a-calgary-energy-company-with-talent-and-a-track-record/

Gjika, G., & Toro, O. (2016). *Albania Oil & Gas Regulation 2016*. Tirana: International Comparative Legal Guides / Gjika & Associates. Retrieved from http://www.iclg.co.uk/practice-areas/oil-and-gas-regulation/oil-and-gas-regulation-2016/albania#chaptercontent2

International Energy Agency. (2008). *Energy in the Western Balkans: The Path to Reform and Reconstruction*. Paris: International Energy Agency. Retrieved from http://www.iea.org/publications/freepublications/publication/balkans2008.pdf

Karburantet, pse nuk pati gjobë nga Konkurrenca? (2015, February 28). *Energjia – Portali i Energjise*. Retrieved from http://energjia.al/2015/02/28/karburantet-pse-nuk-pati-gjobe-nga-konkurrenca/

Kastrati Sh.A. (n.d.). Retrieved from http://kastratigroup.al/

Koleka, B., & Weir, K. (2016, February 2). Petromanas Sells Albanians Assets to Shell for $45mln. *Reuters.com*. Retrieved from http://www.reuters.com/article/albania-petromanas-shell-idUSL8N15H3NN

Kürüm Holding A.Ş. (n.d.). Retrieved from http://www.kurum.com.tr/

Leskoviku, A. (2003, April 10–11). *The Oil Industry in Expanded European Union*. Presentation at the Regional Conference in Portoroz, Slovenia. Retrieved from http://www.world-petroleum.org/docs/docs/pdf/albania_oil_sector.pdf

LISTA E PLOTE / Cilët janë 10 njerëzit më të pasur në Shqipëri. [FULL LIST / Who Are the 10 Richest People in Albania] (2015, July 16). *Panorama Online*. Retrieved from http://www.panorama.com.al/lista-e-plote-cilet-jane-10-njerezit-me-te-pasur-ne-shqiperi/

National Agency of Natural Resources. (n.d.). Retrieved from http://www.akbn.gov.al/

Offshore Company to Reactivate Domestic Oil Refining. (2016, September 1). *Tirana Times*. Retrieved from http://www.tiranatimes.com/?p=128913

Phoenix Petroleum Sh.A. (n.d.). Retrieved from http://phoenixpetroleum.webs.com/

Popovici, V., Deliso, C., Michaletos, I. & Stavros, M. (2012, February 5). Albania Oil Industry Enjoys Revival, But Investor-Government Relations Remain a Question. *Balkanalysis.com Special Report*. Retrieved from http://www.balkanalysis.com/albania/2012/02/05/albania-oil-industry-enjoys-revival-but-investor-government-relations-remain-a-question/

Prapaskenat e kolapsit të gjigandit të hidrokarbureve ARMO. (2016, May 27). *MAPO.* Retrieved from http://www.mapo.al/2016/05/prapaskenat-e-kolapsit-te-gjigandit-te-hidrokarbureve-armo/1

Tirana Court Launches Bankruptcy Procedure Against Albanian Steel Maker Kurum. (2016, April 19). *bne IntelliNews.* Retrieved from http://www.intellinews.com/tirana-court-launches-bankruptcy-procedure-against-albanian-steel-maker-kurum-95546/

TransAtlantic Petroleum Ltd. (n.d.). Retrieved from http://www.transatlanticpetroleum.com/

Transoilgroup AG. (n.d.). Retrieved from http://www.transoilgroup.com/

Vetro Says to Make Deposit Soon to Back Albpetrol Sale. (2012, December 11). *Reuters.com.* Retrieved from http://www.reuters.com/article/albania-albpetrol-sale-idUSL5E8NB8IT20121211

Vlček, T., & Jirušek, M. (2018). The Hydrocarbons Sector in Albania: Short-Term Challenges and Long-Term Opportunities. *Mediterranean Quarterly,* 29(1), 96–119. Durham: Duke University Press.

Will the New Broom in Albania Be Big Enough? (2013, September 24). *bne IntelliNews.* Retrieved from http://www.intellinews.com/will-the-new-broom-in-albania-be-big-enough-500018027/?archive=bne

List of Interviews

Interview 01: Tirana, Albania, April 21, 2016.
Interview 02: Tirana, Albania, April 21, 2016.
Interview 03: Tirana, Albania, April 21, 2016.
Interview 04: Tirana, Albania, April 21, 2016.
Interview 05: Tirana, Albania, April 22, 2016.
Interview 06: Tirana, Albania, April 22, 2016.
Interview 07: Vienna, Austria, July 14, 2016.
Interview 08: Vienna, Austria, May 5, 2017.

CHAPTER 5

Bosnia and Herzegovina

5.1 Crude Oil Sector General Information

5.1.1 Introduction and Upstream

Bosnia and Herzegovina is a Western Balkan country with approximately 3.5 million inhabitants and an area of 51,000 km². It has three neighbours: Croatia to the West and North, Serbia to the East, and Montenegro to the South. Bosnia has limited access to the Adriatic Sea via a 20 km stretch of coastline near the city of Neum.

Under the 1995 Dayton Agreement which ended the war in Bosnia and Herzegovina, the country was divided into two autonomous entities. To the North and East lies *Republika Srpska* (*Republic of Srpska*), whose major ethnic groups are Serb and Bosniak and whose administrative seat is in Banja Luka. The Central and Western parts of the country form the *Federacija Bosne i Hercegovine* (*Federation of Bosnia and Herzegovina*) and are home to Bosniaks and Bosnian Croats as the major ethnic groups, with the administrative seat located in Sarajevo. There is a third entity called the Brčko District, located in the Northeastern part of the country. Since half of the District technically lay in *Republika Srpska* and the other half in *Federacija BiH*, it was not until 1999 that an agreement was reached on how to govern the province. Since that time, it has been a self-governing unit under the shared sovereignty of both the major legal entities in Bosnia and Herzegovina. The president of Bosnia and Herzegovina is elected for

© The Author(s) 2019
T. Vlček, M. Jirušek, *Russian Oil Enterprises in Europe*,
https://doi.org/10.1007/978-3-030-19839-8_5

a four-year term, but three presidents are always elected: a Serb, a Bosniak, and a Croat, who then rotate every eight months in office within their four-year mandate.

The political situation is extremely complex, and ethnic pressures place major limits on cooperation and political development in the country. The relationship between the two major autonomous entities may hardly be labelled cooperative—quite the opposite (Interview 01). This situation makes for a weak state with no real power to enact substantial political and business reforms. Individual deals and agreements closed between one of these entities and a foreign country amount to little more than memoranda, since every bilateral or multilateral agreement must be approved by the central authority, which is usually difficult.

Since 2008, when Rafinerija nafte Brod a.d. was brought back online, Bosnia and Herzegovina restarted crude oil imports (around 1 million tonnes annually [mta]) to supply the facility. The largest share is imported from the Russian Federation (Energy Community Secretariat, 2015, p. 67), purchased from OAO Zarubezhneft. When the refinery was damaged during the civil war in 1991, Bosnia and Herzegovina began to import the oil derivatives that were required. Currently about 700,000 tonnes of derivatives (771 in 2014) are still being imported on top of domestic refining activity, mainly from Croatia, Italy, Hungary, Slovenia, and Austria (Energy Community, 2013, p. 51; Energy Community Secretariat, 2015, p. 67) (Table 5.1).

Oil derivatives constitute 28.5% of final energy consumption in Bosnia and Herzegovina, with 60% consumed in *Federacija BiH* and 40% in *Republika Srpska* (Agency for Statistics of Bosnia and Herzegovina, 2016b, p. 2; Energy Community, 2013, p. 53). The majority is consumed within the transport sector.

Table 5.1 Bosnia and Herzegovina crude oil data

	2008	*2009*	*2012*	*2014*
Consumption	0.120	1.078	0.953	0.951
Production	0	0	0	0
Import dependency	100%	100%	100%	100%

Source: Agency for Statistics of Bosnia and Herzegovina, 2016a, p. 1; Energy Community, 2011, p. 35, 2013, p. 50

Note: Figures in thousands of tonnes; percentage calculations by T. Vlček

Oil exploration and production are at distinct stages in *Federacija BiH* and *Republika Srpska*. There are two promising locations in *Federacija BiH*, the Dinaric Alps that stretch along the Western border with Croatia and the Tuzla region in the Eastern section some 25 km from the Brčko District. Estimates are that the oil deposits in the Tuzla region amount to 50 million tonnes. No oil is currently being extracted in *Federacija BiH*, but some companies have recently expressed serious interest, starting with Royal Dutch Shell in 2011. The company withdrew, however, in 2015 and the bidding process was reopened. Starting in 2015 and in 2016, negotiations have been taking place with Australia's Key Petroleum Limited, France's Total S.A., and the Norwegian firm Spectrum ASA (*"Australia's Key"*, 2015; *"Bosnia Federation"*, 2014; *"Total Eyes"*, 2016; *"Shell Pulls"*, 2015).

Republika Srpska is in a better position as the major oil field in Bosnia and Herzegovina, with estimated reserves of 500 million tonnes of oil located within its borders, specifically in Šamac in the Northeast. Similar to Brod, the city is divided by the border and the Sava River into Slavonski Šamac in Croatia and Šamac in *Republika Srpska*. A 28-year concession was awarded in 2011 by the authority in Banja Luka to Jadran Naftagas d.o.o., a company owned by Russia's AO NeftegazInKor (Neftegazovaja Inovacionnaja Korporacija; 66%) and Serbia's Naftna industrija Srbije A.D. (34%). AO NeftegazInKor is a 100% subsidiary of Russian state-owned OAO Zarubezhneft, and Naftna industrija Srbije A.D.'s major shareholder (56.15%) is PJSC Gazprom Neft, a subsidiary of OAO Gazprom. Drilling started and the first oil was extracted near the village of Obudovac in 2015 (AO NeftegazInKor, n.d.; *"Bosnia Has"*, 2014; *"Joint Venture"*, 2013; Naftna industrija Srbije A.D., n.d.; *"NIS pronašao"*, 2015; OAO Zarubezhneft, n.d.; *"Prva nafta"*, 2014).

Even though some estimates show huge numbers that, if correct, would make *Republika Srpska* the biggest onshore deposit in Europe, a major limitation applies throughout Bosnia and Herzegovina, which is that research and exploration have been limited or completely lacking in promising oil reserve locations. Oil is expected at a fairly deep 6 km under hard limestone beds. Since 2015, there has been no news of any new exploration.

5.1.2   Midstream

Crude oil is imported to Rafinerija nafte Brod a.d. in Brod through the Jadranski Naftovod (JANAF) system[1] from Omišalj, Croatia. The pipeline branch from Slavonski Brod in Croatia to the refinery in Brod is 13-km long. It is owned and operated by Optima Grupa d.o.o., a 100% subsidiary of AO NeftegazInKor. The refinery sits on the right bank of the Sava, in Brod.[2]

5.1.3   Downstream

There is a single crude oil refinery in Brod and one in Modriča. Both refineries are operated by Optima Grupa d.o.o., seated in Banja Luka, where 100% of shares are owned by AO NeftegazInKor. Optima Grupa d.o.o. is, then, the owner of the vast majority of the Bosnia and Herzegovina oil sector. Geographically, the Bosnia and Herzegovina oil sector is located in *Republika Srpska*. Optima Grupa d.o.o. owns 79.998602% of Rafinerija nafte Brod a.d., 75.65% of Rafinerija ulja Modriča a.d., and 80.000031% of Nestro Petrol a.d. Banja Luka. Rafinerija nafte Brod a.d. has a refining capacity of 1.2 mta to date. Rafinerija ulja Modriča a.d., a refinery producing base oils, has a capacity of 90,000 tonnes annually (AO NeftegazInKor, n.d.; Optima Grupa d.o.o., n.d.) (Table 5.2).

Table 5.2   Capacity of Bosnia and Herzegovina refineries as of 2016

Refinery	Owner	Refining capacity	Nelson complexity index (2014)	Year established
Brod	Rafinerija nafte Brod a.d.	4.2 mta	7.1	1968, 1990
Modriča[a]	Rafinerija ulja Modriča a.d.	0.09 mta	–	1954

Source: Compiled by T. Vlček from public sources

[a]Oil refinery producing base oils

[1] See detailed information about the JANAF system in the chapter on Croatia.

[2] The city of Brod is divided by the Sava River, which also forms the border between Croatia and Bosnia and Herzegovina. The Northern (Croatian) section of the city is called Slavonski Brod and the Southern (Bosnia and Herzegovina) section was first called Bosanski Brod, then Srpski Brod during the war, and was eventually renamed Brod with no prefix in 2009.

The refinery in Brod has a long history of operation, having been founded in 1892. The atmospheric distillation unit, today called the 'old line', with a capacity of 1.2 mta, was constructed in 1968. The 'new line' atmospheric distillation unit, with a capacity of 3 mta, was constructed in 1990. The refinery, however, was badly damaged shortly after commissioning during the war in Bosnia and Herzegovina, and the 'old line' did not restart operation until late 2008. Parallel to the operation of the 'old line', reconstruction and repairs on the 'new line' began in 2009 with the goal of increasing nameplate capacity to 4.2 mta.

There are around 365 petrol stations in Bosnia and Herzegovina. The operator of the largest of them is Croatia's INA—Industrija nafte d.d.[3] It owns 100% of HoldINA d.o.o., Sarajevo, which operates 48 petrol stations, and 67% of Energopetrol d.d.,[4] which operates another 55 stations, making up a total of 103 petrol stations (28% market share). Several other major companies also operate in the retail sector of Bosnia and Herzegovina. Nestro Petrol a.d., Banja Luka (80.000031% owned by Optima Grupa d.o.o.) operates a network of 86 petrol stations (74 of which are in *Republika Srpska* and 12 in *Federacija BiH* and the Brčko District), making for a 24% market share. Naftna industrija Srbije A.D. is the 100% owner of G-Petrol d.o.o., Sarajevo and NIS Petrol d.o.o., Banja Luka. In 2012, NIS A.D. purchased OMV BH d.o.o., thereby acquiring 28 petrol stations of the OMV AG subsidiary in the country, 18 of which are situated in *Federacija BiH*. Through G-Petrol d.o.o., Sarajevo, NIS A.D. today operates 36 petrol stations (9.9% market share), 27 of which come under the Gazprom brand. Petrol BH Oil Company d.o.o., a 100% subsidiary of Slovenian Petrol d.d., Ljubljana, operates a network of 35 petrol stations (9.6% market share), the majority of which are located in *Federacija BiH* (*"Bosnia's G-Petrol"*, 2015; Energopetrol d.d., n.d.; *"Ina kupila"*, 2016; *"INA Opened"*, 2016; Nestro Petrol a.d., n.d.). These major players are followed by a number of small companies, some operating only a single petrol station. In terms of Russian state-owned company presence in Bosnia and Herzegovina, OAO Zarubezhneft controls 86 petrol stations (24% market share) and OAO Gazprom (as majority shareholder) 36 petrol stations (9.9% market share).

---

[3] The Hungarian MOL Rt owns 49.08% of the shares in INA—Industrija nafte d.d.

[4] Currently 22.12% of shares are held by *Federacija BiH*, but are also planned for privatization, and INA—Industrija nafte d.d. is interested in becoming the 100% owner of the company (*"Hungary's MOL"*, 2016; *"INA Wants"*, 2016).

5.2   RECENT MARKET DEVELOPMENTS
AND RUSSIAN ACTIVITIES

The ideological notions that underpinned the secession of Croatia and Slovenia from Yugoslavia were also present in Bosnia and Herzegovina and became the basis for the war fought there in 1992–95. The refinery in Brod, which had already ceased operation by 1991, was left with no crude oil because it had always been imported through Croatia, now considered an enemy state. The refinery was seriously damaged during the war. After the signing of the 1995 Dayton Agreement, efforts were made to restart operation of the refinery and this bore fruit in May 2000 when the refinery was reopened. It functioned, however, at only a fraction of its original capacity—15–20% of its pre-war capacity of around 3.2 mta. Bosnia and Herzegovina thus became strongly dependent on oil derivative imports. The Croatian firm INA—Industrija nafte d.d. participated in 1998 in talks with *Republika Srpska* on a potential joint effort to restart the refinery, but they ended with no agreement. Croatia did not, however, abandon its business interests in the region during this time.

In 2002, Croatia limited the export of oil products to Bosnia and Herzegovina to only four border-crossing points. This restriction was aimed at helping the country's own INA d.d. by restricting the transit of cheaper products from Slovenia.[5] Bosnia and Herzegovina was thus forced to purchase more expensive derivatives from INA d.d., and this prompted it to impose sanctions of its own on Croatian products (mainly related to mandatory fuel quality standards) and to focus on the reconstruction and modernization of the Brod refinery as a way of avoiding dependency on foreign products (Latal, 2002).

The major limitation, however, was a lack of investment. The refinery struggled to survive under limited operation and tight funding, piled up debts, and was eventually forced to shut down in 2005. This caused serious unemployment in Brod and the region struggled with a major economic setback. In 2007, after two years of negotiation, *Republika Srpska* closed a direct deal without public procurement procedure (criticized by some local non-governmental organizations [NGOs] and the media) with

---

[5] The eventual restart of the Brod Refinery in 2008 was against Croatia's interests, and Croatia tried to secure a better environment for INA d.d. by pricing OAO Zarubezhneft's crude oil imports to Brod higher than INA d.d.'s imports to Croatia. The Russian Embassies in Sarajevo and Zagreb were a mitigating influence, and the Russian Embassy in Croatia negotiated a lower price (Tepavcevic, 2015, p. 47).

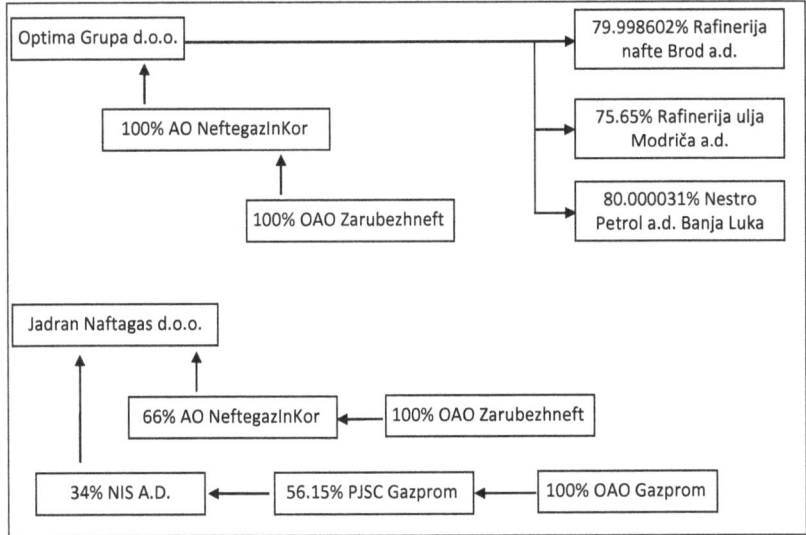

Fig. 5.1 Diagram of Russian state companies' ownership of Optima Grupa d.o.o. and Jadran Naftagas d.o.o. Source: T. Vlček. Note: Situation as of 2016

the state-owned Russian firm OAO Zarubezhneft. This created an opportunity for AO NeftegazInKor, seated in Banja Luka in *Republika Srpska*. AO NeftegazInKor paid a total of EUR 125.8 million for a 75% share in the Brod refinery, a 66.75% share in the Modriča oil refinery, a 70% share in Nestro Petrol a.d., and pledged to repay the debts of these three companies (exceeding EUR 72 million) and invest a further EUR 600–700 million in modernizing *Republika Srpska*'s oil industry (*"Russia-owned"*, 2008). The acquisition was carried out via the purchase of 100% of the shares of the owner, Optima Grupa d.o.o. (see Fig. 5.1).

But it would be erroneous to presume that OAO Zarubezhneft's entry into *Republika Srpska* was in the Russian state interest. The initiative came from the *Republika Srpska* diaspora in the Russian Federation. The state-owned company made an independent business decision and actually had to struggle through complicated negotiations with the Kremlin for the state credit needed to make the acquisition (Tepavcevic, 2015, p. 46, 48). According to a former OAO Zarubezhneft official, the main motive for investing in the oil capacities of Bosnia and Herzegovina was the easier access it would provide the company to customers in the EU markets

(Tepavcevic, 2015, p. 44). The company was eventually able to persuade the officials and received a EUR 350 million credit from VneshEconomBank (Russian Development Bank) (*"Russia Kindles"*, 2009). The investment did have political effects: it further strengthened the ties that bound Serbs (*Republika Srpska*) in Bosnia and Herzegovina to Russia, for the effects of the investment brought economic revival to cities and regions, saw the return of enterprises to *Republika Srpska*, and strengthened the country's economic performance (Optima Grupa d.o.o. is a sizeable economic entity and a major contributor to the state budget in Bosnia and Herzegovina) and political power (Pivovarenko, 2014).

Under OAO Zarubezhneft's ownership, the refinery restarted operations in 2008. Repair and reconstruction of the 'new line' have been under way since 2009 to increase nameplate capacity to 4.2 mta. Since 2008, the refinery has processed Russian crude oil imported via the JANAF pipeline. The crude oil is purchased from OAO Zarubezhneft. But reconstruction may potentially take much longer than planned because of the problematic financial performance of Optima Grupa. Although Aleksandar Dijakonov, the company CEO, expected the refinery to become profitable in 2015, the reality falls far short of that. Optima Grupa d.o.o. has accumulated considerable losses—more than EUR 430 million—since 2010 (*"Bosnia's Nestro"*, 2015; *"Bosnia's Optima"*, 2015; *"Continuation of "*, 2013). These losses are said to be higher than the company's equity value (Interview 07). Around 2015–16, OAO Zarubezhneft began to consider withdrawing from Bosnia and Herzegovina, and Kuwait expressed interest in taking over the Brod refinery should the current owner be willing to sell (*"Kuwait Interested"*, 2016). At the moment, though, there are no takers: the primary reason for Optima Grupa's poor financial condition is the alleged syphoning of company funds through the Virgin Islands-based Parthenon company (Interview 07).

Naftna industrija Srbije A.D. (and thus OAO Gazprom) entered the Bosnia and Herzegovina market in 2010 and 2012. In 2010, the firm Jadran Naftagas d.o.o. was established. Owned by AO NeftegazInKor (66%) and NIS A.D. (33%), it was awarded a 28-year concession (3 years for exploration and 25 years for exploitation) by the *Republika Srpska* authority in Banja Luka (Ministry of Industry, Energy, and Mining). Two Russian state-owned enterprises, OAO Zarubezhneft and OAO Gazprom, are thus shareholders of Jadran Naftagas d.o.o., which committed to investing US $40.7 million in exploration and US $188.3 million in exploitation if reserves were proven in Šamac, *Republika Srpska* (*"Jadran*

Naftagas", 2012; "*Total Eyes*", 2016). Drilling started and the first oil was extracted near the village of Obudovac in 2015.

In 2012, Naftna industrija Srbije A.D. purchased OMV BH d.o.o for an undisclosed sum, thereby acquiring the 28 petrol stations owned by the OMV AG subsidiary in Bosnia and Herzegovina, 18 of which lay in *Federacija BiH*. This was in line with NIS A.D.'s strategy of expanding its retail network across the countries of the Balkan region; the Bosnian and Herzegovinian market, together with the Bulgarian and Romanian markets, is a key target in the company's external strategy for broadening its retail network with Euro 5 quality products ("*Naftna industrija*", 2012; "*NIS Acquires*", 2012). Rumours have it that OAO Gazprom petrol stations are slashing their product prices to attract customers (Interview 04). Recently, the company announced its intention to sell three petrol stations in Gračanica, Špionica, and Tinja ("*NIS to Sell*", 2015; "*Serbia's NIS*", 2015). All three villages are in *Federacija BiH*, which raises the possibility of the business environment in *Federacija BiH* becoming complicated for Russian companies.

At the same time, it should be noted that the relationship between the Russian Federation and *Republika Srpska* is often exaggerated. For example, the entire annual budget of *Republika Srpska* is lower than the annual budget of the canton of Sarajevo. Russian influence is exaggerated by the media; the Russian strategy is to dominate the media space and supply the audience with information that will reinforce its imagination (Interview 03). It is also important to note that there is basically no independent media. All media channels, newspapers, websites, and TV channels are financed by subjects with vested interests, be they local or foreign business people, ethnic groups, political parties, or foreign countries (Interview 01; Interview 08).

5.3   Research Indicator Assessment

5.3.1   *Active Support by Russian State Representatives for Energy Enterprises and Their Activities Abroad*

The entry of OAO Zarubezhneft into the Bosnian and Herzegovinian oil market can hardly be described as part of a Kremlin strategy. The origins of its entrance lie in the disputes between Bosnia and Herzegovina and Croatia, especially the 2002 dispute (see earlier). Bosnia and Herzegovina has focused strongly on the reconstruction and modernization of the Brod refinery since 2002 to avoid dependency on products from abroad. The search for an investor was, however, very difficult, and activity by the *Republika Srpska* diaspora in the Russian Federation was also a factor. Members eventually succeeded in motivating OAO Zarubezhneft to act, but the company later struggled with the Kremlin for the state credit needed to make the acquisition. This process can hardly be ascribed to strategic behaviour on the part of Russian state-owned enterprises (SOEs). Rather, it was tied to OAO Zarubezhneft's easier access to customers in EU markets and the interest of *Republika Srpska* representatives in their struggle with Croatia.

On the other hand, when the acquisition was completed, there were clear signs of the Kremlin's support for the enterprise, as in the 2008 dispute over the price of oil transport. The eventual restart of the Brod refinery in 2008 ran counter to Croatia's interests and Croatia tried to secure a better environment for INA d.d. by pricing OAO Zarubezhneft's crude oil imports to Brod higher than INA d.d.'s imports to Croatia. This was mitigated by the involvement of the Russian Embassies in Sarajevo and Zagreb, and the Russian Embassy in Croatia negotiated a lower price (Tepavcevic, 2015, p. 47).

The operations of Naftna industrija Srbije A.D. in Bosnia and Herzegovina are rather limited (9.9% retail market share) and market-related as they are part of NIS A.D.'s regional asset development strategy.

5.3.2   As a Foreign Supplier, Russia Rewards Certain Behaviours and Links Energy Deals to the Client State's Foreign Policy Orientation

Russian operations are basically limited to *Republika Srpska*. There are many reasons for this, including the Slavonic ethnicity of the majority (Bosnian Serbs), the common Orthodox religion, historical relations between Russia and Serbia and Pan-Slavism, Russian-peacekeeping troops deployed in the UNPROFOR, IFOR, and SFOR missions in Croatia and Bosnia and Herzegovina,[6] the Russian position on postwar reconstruction, and good relations with the president of *Republika Srpska*.

---

[6]The Russian Federation withdrew all troops from these missions and froze all NATO-Russia military and political cooperation after the NATO air campaign in Serbia in 1999 (Nikitin, 2004).

However, the current Russian relationship with *Republika Srpska* originated in the wake of the strong political leadership of President Milorad Dodik, who was outstanding among Serb political leaders in the region, and therefore supported (Gajic, 2015). Russian support takes different forms, including, for example, meetings with officials (as when Milorad Dodik met Vladimir Putin in Moscow in September 2014 shortly before the elections in *Republika Srpska*) or Russian loans to *Republika Srpska*. The reality of any loans announced, though, is often quite different from what is promised. For example, the 2014 EUR 270 million loan promised by Moscow (offered under better conditions than those of the International Monetary Fund [IMF] loan negotiated at that time) was never actually made. Some say the promise was just used by Milorad Dodik to gain points in the election campaign (*"BN TV"*, 2015). The same charges followed the devastating floods of 2014, when Milorad Dodik also announced a Russian loan of EUR 500–700 million to help stabilize finances after the disaster (*"Bosnian Serb"*, 2014). This loan was also never issued, and Bosnia and Herzegovina eventually secured an IMF loan for EUR 550 million in May 2016 (Koseva, 2016). This loan was aimed at supporting the country's economic reform agenda. The first tranche (of approximately EUR 90 million) was paid by the IMF with no conditions imposed on Bosnia and Herzegovina (Interview 01). But it was put on hold in April 2017 because the country was not able to push through the required reforms (*"IMF Delays"*, 2017). Milorad Dodik is nevertheless serving his second term in office, having been re-elected in October 2014. Dodik's primary target is capitalization, as he is somehow involved in almost every business in *Republika Srpska* (Interview 03).

Also, the 2017 decision by Russia to repay the Soviet-era debt of US $125 million to Bosnia and Herzegovina (divided between the three entities: 58% to *Federacija BiH*, 29% to *Republika Srpska*, and 10% to Brčko District; *"Russia to Repay"*, 2017) was part of Russia's image-making efforts: the timing of the announcement is noteworthy given Russia's loss of influence in Montenegro. Bosnia and Herzegovina could possibly become a new base for the Russian presence in the Balkan Peninsula (Interview 01). This debt repayment is also likely connected to the problems surrounding the IMF loan, from which the Russian Federation can earn political capital (Interview 04) since Russian investments generally always aim at earning public support by funding visible endeavours such as schools, infrastructure, and so on.

Bosnia and Herzegovina and *Republika Srpska* are, however, not vital trajectories within Russia's foreign policy. Russian support has tended to be verbal, without much tangible benefit for Bosnia and Herzegovina. The good relations with *Republika Srpska* are more the outcome of Russia's traditional support and protection for Serbs and other Orthodox Slavs in South-Eastern Europe. There is, however, the question of Russia's potential power to destabilize the Southern border of the EU by stirring up the historical mutual intolerance between Bosnian Serbs and Croats. The Russian Federation has this power (through the UN Security Council, through the Russian deputy head of the Organization for Security and Co-operation in Europe [OSCE] mission to Bosnia and Herzegovina, through Russia's permanent representation in Partnership for Peace, etc.) and uses it to block the country's efforts to enter the European Union and maintain the status quo in Bosnia and Herzegovina. Thus, Russia sometimes uses the threat of a veto in the UN Security Council against the UN/EU mission EUFOR ALTHEA, which would cause the mission to lose its legal mandate (Interview 05).

5.3.3   Abuse of Infrastructure (e.g. Pipelines) and Differential Pricing to Exert Pressure on the Client State

No indications were found in the oil sector.

5.3.4   Efforts to Take Control of the Energy Resources, Transit Routes, and Distribution Networks of the Client State

Other than Jadran Naftagas d.o.o. already controlling the deposit in Šamac via the 28-year concession awarded in 2011, no indications were found of Russian SOE interest in controlling infrastructure. The 2012 Naftna industrija Srbije A.D. purchase of OMV BH d.o.o.'s 28 petrol stations is part of the company's external market strategy, not an effort to control distribution.

5.3.5   Disruption (by Various Means) of Alternative Supply Routes/Sources of Supply

No indications were found in the oil sector.

5.3.6   Efforts to Gain a Dominant Market Position in the Client Country

Such efforts were indeed made, especially in connection with OAO Zarubezhneft's 2008 entry into the Bosnia and Herzegovina oil market. However, as stated earlier, these efforts can hardly be ascribed to strategic considerations by the Kremlin. Rather they are part of the business plans of OAO Zarubezhneft. Although the company is state-owned, it still had to persuade the Kremlin of the logic of the planned acquisition of Optima Grupa d.o.o. in very complicated discussions.

5.3.7   Efforts to Eliminate Competitive Suppliers

No indications were found in the oil sector.

5.3.8   Acting Against Liberalization

No indications were found in the oil sector.

5.3.9   Diminishing the Importance and Influence of Multilateral Regimes Such as the EU

The legal system and market rules do not follow EU rules, since the country is not a member state. Bosnia and Herzegovina takes part in the Stabilisation and Association Process and formally applied for EU membership in February 2016, but is still far from accession. Multilateral regimes therefore do not play a significant role in Russia-Bosnia Herzegovina relations. When negotiating individual projects, Russia does not approach the Office of the High Representative as the principal arm of government, but negotiates directly with the autonomous entities, especially with *Republika Srpska*.

5.3.10   Attempts to Control the Entire Supply Chain (Regardless of Commercial Rationale)

No indications were found in the oil sector.

5.3.11   Economically Irrational Steps Taken to Maintain a Particular Position in the Client State's Market

No indications were found in the oil sector.

Table 5.3 Summary of indicators

Indicator	Found	Found with
Active support by Russian state representatives for energy enterprises and their activities abroad	No	–
As a foreign supplier, Russia rewards certain behaviours and links energy deals to the client state's foreign policy orientation	Inconclusive	–
Abuse of infrastructure (e.g. pipelines) and differential pricing to exert pressure on the client state	No	–
Efforts to take control of the energy resources, transit routes, and distribution networks of the client state	No	–
Disruption (by various means) of alternative supply routes/sources of supply	No	–
Efforts to gain a dominant market position in the client country	Yes	OAO Zarubezhneft
Efforts to eliminate competitive suppliers	Inconclusive	–
Acting against liberalization	No	–
Diminishing the importance and influence of multilateral regimes such as the EU	Yes	–
Attempts to control the entire supply chain (regardless of commercial rationale)	No	–
Economically irrational steps taken to maintain a particular position in the client state's market	No	–

Source: Author

Note: *Inconclusive* means some indications of this behaviour were found, but not of a shape, size, or importance to be ascribed to strategic behaviour and thus fulfil the indicator

SOURCES

Agency for Statistics of Bosnia and Herzegovina. (2016a). *Energy Statistics Oil, Petroleum Products, 2014.* Retrieved from http://www.bhas.ba/saop-stenja/2016/END_2014G01_001_01_BS.pdf

Agency for Statistics of Bosnia and Herzegovina. (2016b). *Energy Statistics Total Energy Balance, BiH, 2014.* Retrieved from http://www.bhas.ba/saop-stenja/2016/ENB_2014G01_001_01_BS.pdf

AO NeftegazInKor. (n.d.). Retrieved from http://www.neftegazincor.ru/

Australia's Key Petroleum Eyes Oil Exploration in Bosnian Region. (2015, December 17). *Reuters.* Retrieved from http://af.reuters.com/article/energyOilNews/idAFL8N14628P20151217

BN TV: Mysterious Russian Loan to RS Will Not Be Paid. (2015, May 3). *Bosnia Today.* Retrieved from http://www.bosniatoday.ba/bn-tv-mysterious-russian-loan-to-rs-will-not-be-paid/

Bosnia 'Has Huge Oil Deposits'. (2014, August 22). *Balkan Insight*. Retrieved from http://www.balkaninsight.com/en/article/bosnia-has-huge-oil-deposits

Bosnia Federation Eyes 2015 Start to Oil Digs. (2014, August 22). *Balkan Insight*. Retrieved October 13, 2016, from http://www.balkaninsight.com/en/article/bosnia-lays-hopes-on-oil-explorations

Bosnia's G-Petrol Launches First NIS Filling Station in Eastern Herzegovina – Report. (2015, July 9). *SeeNews*. Retrieved from https://seenews.com/news/bosnias-g-petrol-launches-first-nis-filling-station-in-eastern-herzegovina-report-483596

Bosnia's Nestro Petrol Plans to Expand Network by Opening Mini Fuel Stations. (2015, July 30). *SeeNews*. Retrieved from https://seenews.com/news/bosnias-nestro-petrol-plans-to-expand-network-by-opening-mini-fuel-stations-486459

Bosnia's Optima Grupa Accumulates 434.6 mln Euro in Losses Over Past 5 yrs – Report. (2015, July 24). *SeeNews*. Retrieved from https://seenews.com/news/bosnias-optima-grupa-accumulates-434-6-mln-euro-in-losses-over-past-5-yrs-report-485586

Bosnian Serb Leader Says Russia Will Loan Region 500–700 mln Euros. (2014, September 19). *Reuters*. Retrieved from http://www.reuters.com/article/bosnia-russia-loan-idUSL6N0RK2XL20140919

Continuation of Investment in "Oil Refinery" in Brod. (2013, April 10). *SarajevoTimes*. Retrieved from http://www.sarajevotimes.com/?p=20680

Energopetrol d.d. (n.d.). Retrieved from http://www.energopetrol.ba/

Energy Community. (2011). *Statement on Security of Energy Supply of Bosnia and Herzegovina*. Retrieved from https://www.energy-community.org/portal/page/portal/ENC_HOME/DOCS/1218177/0633975AB59F7B9CE053C92FA8C06338.PDF

Energy Community. (2013). *Statement on Security of Energy Supply of Bosnia and Herzegovina*. Retrieved from https://www.energy-community.org/portal/page/portal/ENC_HOME/DOCS/2422180/0633975AD43E7B9CE053C92FA8C06338.PDF

Energy Community Secretariat. (2015). *Annual Implementation Report 2014/2015*. Vienna: Energy Community Secretariat.

Gajic, S. (2015, July 30). Why Is Bosnia Important to Russia? *Fort Russ*. Retrieved from http://www.fort-russ.com/2015/07/why-is-bosnia-important-to-russia.html

Hungary's MOL Exits Bosnia's Energopetrol, Croatia's INA Raises Stake to 67%. (2016, July 12). *SeeNews*. Retrieved from https://seenews.com/news/hungarys-mol-exits-bosnias-energopetrol-croatias-ina-raises-stake-to-67-532273

IMF Delays Cash After Bosnia MPs Block Reform. (2017, April 6). *Balkan Insight*. Retrieved from http://www.balkaninsight.com/en/article/imf-loan-hit-by-delays-after-bosnian-mps-fail-on-reforms-04-06-2017-1

Ina kupila 33,5 posto Molovih dionica Energopetrola. (2016, July 12). *Poslovni denik*. Retrieved from http://www.poslovni.hr/domace-kompanije/ina-kupila-335-posto-molovih-dionica-energopetrola-315307

INA Opened Two New Petrol Stations in BiH – Bosnia and Herzegovina. (2016, June 9). *Balkan Energy*. Retrieved from http://balkanenergy.com/ina-opened-two-new-petrol-stations-in-bih-bosnia-and-herzegovina-9-june-2016/

INA Wants to Buy FBiH's Share in Energopetrol – Bosnia and Herzegovina. (2016, July 14). *Balkan Energy*. Retrieved from http://balkanenergy.com/ina-wants-to-buy-fbihs-share-in-energopetrol-bosnia-and-herzegovina-14-july-2016/

Jadran Naftagas Presented Preliminary Results of Oil and Gas Exploration at the Territory of the Republic of Srpska. (2012, September 26). *NIS A.D. News*. Retrieved from http://ir.nis.eu/news-and-events/single-news/article/330/

Joint Venture with Serbian NIS Spuds First Exploration Well in Bosnia and Herzegovina. (2013, June 4). *Gazprom Neft Newsroom*. Retrieved from http://www.gazprom-neft.com/press-center/news/1095057/

Koseva, D. (2016, May 25). IMF, Bosnia Agree New €550mn Loan Deal. *Bne IntelliNews*. Retrieved from http://www.intellinews.com/imf-bosnia-agree-new-550mn-loan-deal-98205/

Kuwait Interested in Buying Bosanski Brod Refinery. (2016, January 18). *Bosnia Today*. Retrieved from http://www.bosniatoday.ba/kuwait-interested-in-buying-bosanski-brod-refinery/

Latal, Z. (2002, February 25). Balkan Oil War. *Global Policy Forum*. Retrieved from https://www.globalpolicy.org/component/content/article/198/40161.html

Naftna industrija Srbije A.D. (n.d.). Retrieved from http://www.nis.eu/

Naftna industrija Srbije kupila pumpe OMV BH. (2012, November 30). *Aljazeera Balkans*. Retrieved from http://balkans.aljazeera.net/vijesti/naftna-industrija-srbije-kupila-pumpe-omv-bh

Nestro Petrol a.d. (n.d.). Retrieved from http://www.nestropetrol.com/

Nikitin, A. (2004). Partners in Peacekeeping. *NATO Review*, 4. Retrieved from http://www.nato.int/docu/review/2004/issue4/english/special.html

NIS Acquires OMV Petrol Stations in BiH. (2012, November 30). *Gazprom Neft News*. Retrieved from http://www.gazprom-neft.com/press-center/news/841515/

NIS pronašao naftu u Republici Srpskoj. (2015, February 23). *N1*. Retrieved from http://rs.n1info.com/a37500/Biznis/NIS-pronasao-naftu-u-Republici-Srpskoj.html

NIS to Sell 3 Petrol Stations in Bosnia. (2015, August 26). *WiseBroker*. Retrieved from http://wisebroker.rs/niis-nis-to-sell-3-petrol-stations-in-bosnia/

OAO Zarubezhneft. (n.d.). Retrieved from http://www.zarubezhneft.ru/

Optima Grupa d.o.o. (n.d.). Retrieved from http://optimagrupa.net/

Pivovarenko, A. (2014, May 23). Modern Russia in the Modern Balkans: Soft Power Through Investment. *Russian International Affairs Council.* Retrieved from http://russiancouncil.ru/en/inner/?id_4=3744

Prva naftna bušotina na teritoriji BiH. (2014, June 6). *Aljazeera Balkans.* Retrieved from http://balkans.aljazeera.net/vijesti/prva-naftna-busotina-na-teritoriji-bih

Russia Kindles Flame of Hope in Bosnia Refinery. (2009, January 4). *Reuters.* Retrieved from http://www.reuters.com/article/us-balkans-russia-idUS-TRE50400H20090105

Russia to Repay Soviet-Era Debt to Bosnia. (2017, August 8). *Balkan Insight.* Retrieved from http://www.balkaninsight.com/en/article/russia-repays-bosnia-old-soviet-debt-08-08-2017

Russia-owned Bosnian Oil Refinery Reopens. (2008, November 27). *Balkan Insight.* Retrieved from http://www.balkaninsight.com/en/article/russia-owned-bosnian-oil-refinery-reopens

Serbia's NIS Lists for Sale Three Filling Stations in Bosnia – Report. (2015, August 25). *SeeNews.* Retrieved from https://seenews.com/news/serbias-nis-lists-for-sale-three-filling-stations-in-bosnia-report-489959

Shell Pulls Out of Bosnia's Quest for Oil. (2015, September 30). *Reuters.* Retrieved from http://af.reuters.com/article/energyOilNews/idAFL5N120365201 50930

Tepavcevic, S. (2015). The Motives of Russian State-Owned Companies for Outward Foreign Direct Investment and Its Impact on State-Company Cooperation: Observations Concerning the Energy Sector. *Transnational Corporations, 23*(1), 29–58. Retrieved from http://unctad.org/en/PublicationChapters/diaeia2015d1a2_en.pdf

Total Eyes Oil Exploration in Bosnia. (2016, March 17). *Newsbase.* Retrieved from http://newsbase.com/topstories/total-eyes-oil-exploration-bosnia

LIST OF INTERVIEWS

Interview 01: Sarajevo, Bosnia and Herzegovina, April 24, 2017.
Interview 02: Sarajevo, Bosnia and Herzegovina, April 24, 2017.
Interview 03: Sarajevo, Bosnia and Herzegovina, April 25, 2017.
Interview 04: Sarajevo, Bosnia and Herzegovina, April 25, 2017.
Interview 05: Sarajevo, Bosnia and Herzegovina, April 25, 2017.
Interview 06: Sarajevo, Bosnia and Herzegovina, April 26, 2017.
Interview 07: Sarajevo, Bosnia and Herzegovina, April 26, 2017.
Interview 08: Sarajevo, Bosnia and Herzegovina, April 26, 2017.

CHAPTER 6

Bulgaria

6.1   Crude Oil Sector General Information

6.1.1   Introduction and Upstream

Bulgaria is an EU member country located in South-Eastern Europe neighbouring two successor countries of the former Yugoslavia (Serbia and Former Yugoslav Republic of Macedonia [FYROM]), as well as Greece, Romania, and Turkey. Bulgarian crude oil consumption is not particularly high; it amounts to around 3.8 million tonnes annually. There are nevertheless imports of roughly double that value (around 7.5 mta; Center for the Study of Democracy, 2014, p. 65; Nitzov, Stefanov, Nikolova, & Hristov, 2010, p. 1) because Bulgaria produces oil products in the Lukoil Neftochim Burgas AD refinery, the largest in the Balkans, with a capacity of 9.5 mta. Crude oil is a dominant energy source, accounting for around 39% of the Bulgarian TPES in 2013 (International Energy Agency, n.d.).

Crude oil imports come to Bulgaria primarily from two sources, the vast majority from the Russian Federation (over 80%) and about 15% from

---

This chapter partially builds on and develops a paper entitled 'Russia's Energy Relations in Southeastern Europe: An Analysis of Motives in Bulgaria and Greece' (Jirušek, Vlček, & Henderson, 2017), published in the Taylor & Francis journal *Post-Soviet Affairs* (www.tandfonline.com).

© The Author(s) 2019
T. Vlček, M. Jirušek, *Russian Oil Enterprises in Europe*,
https://doi.org/10.1007/978-3-030-19839-8_6

Table 6.1   Bulgaria crude oil data

	2010	2012	2013	2014
Consumption	3900	3900	3600	3800
Production	23	24	28	n/a
Import dependency	100%	100%	100%	100%

Source: BP plc, 2015, p. 11; International Energy Agency, n.d.

Note: Figures in thousands of tonnes annually

Kazakhstan via the Caspian Pipeline Consortium (CPC) pipeline (Center for the Study of Democracy, 2014, p. 65; Stankova, n.d.). At approximately 3 million tonnes annually, this makes Russia the key supplier of crude oil. This is attributable to the main crude oil importer in Bulgaria being the Lukoil-owned Neftochim Burgas refinery (Table 6.1).

Bulgaria's domestic crude oil production is tiny, between 22,000 and 28,000 tonnes of oil annually on average over the long term, which makes the country totally dependent on imports. All current production takes place near Pleven in the North of the country. Oil and Gas Exploration and Production Plc (OGEP) has concessions for the extraction of crude oil from more than ten oil fields in Gorni Dabnik, Dolni Lukovit, Staroseltzi, Bardarski geran, Selanovtzi, and Tyulenovo (Chimimport Plc, 2014, p. 23). There are several survey projects in Northern Bulgaria and especially along the Black Sea coast in which foreign companies also participate (Dutch Shell, Ireland's Petroceltic, France's Total, Austria's OMV, Repsol from Spain, and others).

The only Bulgarian company active in oil survey and exploration is Oil and Gas Exploration and Production Plc (OGEP), created in 1991 as a state enterprise. The company was privatized in 2003 and the majority owner is Chimimport Plc (55.91%) through Zarneni Hrani Bulgaria AD[1] (Chimimport Plc, 2013, p. 19). OGEP also operates in the retail market via Chimoil BG Ltd.

6.1.2   Midstream

When it comes to infrastructure, there is but a single oil pipeline in Bulgaria, and it runs from Tyulenovo (on the Black Sea shore in the Northeast of the country) to Pleven (in Northern Bulgaria), where the

---

[1] Chimimport Plc owns 63.68% of shares in Zarneni Hrani Bulgaria AD (Chimimport Plc, 2013, p. 19).

Table 6.2 Capacity of Bulgarian refineries as of 2013

Refinery	Owner	Refining capacity	Type	Year established
Neftochim Burgas	Lukoil Neftochim Burgas AD (≈100% Lukoil Europe Holdings B.V.)	9.5 mta	Cracking (FCC), Hydrocracking, vacuum and atmospheric distillation	1963
Sofia Petroleum Refinery	Bulgarian Petroleum Refinery Ltd. (100% Oil and Gas Exploration and Production Plc)	–	Atmospheric distillation	1994

Source: Bulgarian Petroleum Refinery Ltd., n.d.; Lukoil Neftochim Burgas AD, n.d.; ЛУКОЙЛ Нефтохим Бургас АД, 2014, p. 8

Plama refinery formerly stood. The refinery has been out of operation since 1998 and the pipeline is no longer used. Crude oil imports thus arrive exclusively by sea, mainly through the Burgas oil terminal Rosenets (owned by Lukoil Neftochim Burgas AD), and partially also through the Varna oil terminal (owned by Oiltanking Bulgaria AD[2]). From Rosenets, crude oil is transported to the Neftochim Burgas refinery via a 15 km pipeline owned by the refinery.

6.1.3 Downstream

There are two refineries in Bulgaria: one, the OGEP-owned Sofia refinery, is of miniscule capacity and has negligible market share. Besides the Neftochim in Burgas on the Black Sea coast—the largest refinery in the Balkan region—two other facilities in Bulgaria were historically located in Pleven (the Plama refinery) and Ruse. Neither is functional today. The Neftochim refinery is under the ownership of Lukoil Neftochim Burgas AD, a company owned by Lukoil Europe Holdings B.V., a subsidiary of Russia's PAO Lukoil (Table 6.2).

LITASCO SA (PAO Lukoil's 100% subsidiary) manages crude oil and other feedstock purchases and deliveries for PAO Lukoil's refineries in Bulgaria, Italy, the Netherlands, and Romania. On behalf of Lukoil's refining and marketing operations in Eastern Europe, LITASCO SA also manages feedstock supply and product sales out of these assets (LITASCO SA, n.d.).

---

[2] Owned by Hamburg-based Marquard & Bahls AG.

The greatest amount of oil is consumed in transport (66%), followed by industry (17%), residential services, agriculture, power plants, and others (2011 data; Enerdata, 2013, p. 20). The oil products' retail market has been thoroughly liberalized, with many players including Petrol Holding AD (operating over 500 petrol stations through its subsidiaries Petrol AD, Naftex Petrol EOOD, and Petrol Gas OOD), Lukoil Bulgaria Ltd. (over 220 petrol stations), Shell Bulgaria EAD (over 110 petrol stations), OMV Bulgaria OOD (95 petrol stations), EKO Bulgaria EAD[3] (81 petrol stations), Prista Oil Holding EAD, Rompetrol Bulgaria EAD (61 petrol stations), NIS EOOD (35 petrol stations), Chimoil BG Ltd (30 petrol stations), and others (Ministry of Economy of the Republic of Bulgaria, 2012, p. 8). A substantial proportion of the total number of around 3190 retail petrol stations were in the hands of small owners as of 2014 (Ministry of Environment and Water of the Republic of Bulgaria, 2014, p. 100).

Russian capital is present in Lukoil Bulgaria Ltd., a subsidiary of the PAO Lukoil operating network of over 220 petrol stations in Bulgaria (around 6.9% of the market), and NIS Petrol EOOD (daughter of NIS Gazprom Neft[4]), which operates a network of 35 Gazprom petrol stations in Bulgaria (around 1.1% of the market). A Russian citizen is also a shareholder in Petrol Holding AD, the biggest retail network operator, with 52.5% of shares owned by Kirsan Ilyumzhinov, the former president of the Republic of Kalmykia, a Southwestern federal entity in the Russian Federation,[5] through his Switzerland-based Credit Mediterranee.

In terms of exports, Bulgarian petroleum products are transported mainly to neighbouring countries, with the crucial markets being Romania and Greece (which together import over 90% of the Lukoil Neftochim Burgas AD refinery's products).

6.2   Recent Market Developments and Russian Activity

Regarding Russian activity in the Bulgarian oil sector, the privatization of the Neftochim refinery and Petrol AD in the 1990s and 2000s will be noted, along with information on the Albanian Macedonian Bulgarian Oil (AMBO) pipeline project—one of the alternatives for bypassing the Turkish Straits.

---

[3] Owned by Greek Hellenic Petroleum S.A.
[4] 56.15% owned by OAO Gazprom.
[5] The remaining 47.5% is owned by Bulgarian businessman Mitko Sabev (Angelov, 2012).

The Neftochim refinery was established in 1963, primarily to refine the Russian Export Blend (REB) due to the political situation. The technology used was thus adapted to the imported crude oil, and using REB it operates most economically, with the best utilization and unit costs. It is thus not surprising that the refinery historically imported about 80% of its needs from the former USSR, and this continues to be the case today. Smaller exporters to Bulgaria like Kazakhstan and Iran do offer a similar crude oil type, heavy and sulphurous, that can be used interchangeably.

In 1997, when the Union of Democratic Forces defeated the Bulgarian Socialist Party—a revamped version of the Bulgarian Communist Party—at the ballot box, the ensuing political changes allowed the Bulgarian Privatization Agency to privatize the country's oil sector. The Neftochim refinery, strategically positioned close to the Burgas maritime port and connected by a 15 km pipeline to the Rosenets oil terminal, contributed over a quarter of the revenue in the national budget and even today accounts for about 25% of public revenues from excise duties and taxes in Bulgaria (Center for the Study of Democracy, 2014, p. 66; Ivanova & Tsolova, 2011; Vatansever, 2006, p. 18). This makes the refinery a highly attractive asset and affords the owner considerable leverage over Bulgaria's national economy.

Because crude oil supply contracts were not put out to tender, the refinery accrued considerable losses and ran up debts to crude oil delivery companies in the 1990s; US $100 million by May 1997 to OAO NK Rosneft alone (Platts Oilgram News, 1997, cited in Vatansever, 2006, p. 19). To avoid monopolization of the Bulgarian oil sector, the privatization of the Neftochim refinery and Petrol AD were treated separately. After two rounds of offers by bidders in 1999, the Bulgarian Privatization Agency eventually selected Russia's PAO Lukoil over the Balkan Petroleum Consortium (Yukos Petroleum Bulgaria, a subsidiary of OAO NK Yukos, Petrol Holding Group AD, OAO NK Rosneft, OAO NGK Slavneft, OAO AK Transneft, OAO Stroytransgaz, ZAO Korporatsiya Orelneft, and Yuden Technology Ltd.), Logomat Services Ltd. (a Cyprus-based company that was dissolved in 2014), and the Turkish firm Akmaya Sanayi Ve Ticaret A.Ş. (disqualified for failing to submit its offer on time) (Vatansever, 2006, pp. 19–21). PAO Lukoil won the tender for a 58% stake in the refinery at a cost of US $100 million, investments of US $400 million in modernizing the refinery complex, and agreeing to take on all of Neftochim's US $230 million debt (Zashev, 2006, p. 118).

One of the remarkable aspects of the privatization process was that Russian companies PAO Lukoil and Yukos Petroleum Bulgaria competed against each other, bidding up the price of the asset. Adnan Vatansever of King's College London adds that support from the Russian government was strong but uncoordinated and shifted from supporting the Yukos Petroleum Bulgaria-led consortium at the start to PAO Lukoil later in 1999. This has been attributed to the fact that OAO NK Rosneft and OAO NGK Slavneft were allegedly created by the Russian fuel and energy ministry to boost deliveries to their own refineries in the Russian Federation; rumour also has it that the change was affected by PAO Lukoil's successful lobbying of the Russian fuel and energy ministry (Vatansever, 2006, p. 21, 30).

Lukoil Europe Holdings B.V. originally purchased 58% of the refinery as a result of the privatization process by the Bulgarian Privatization Agency in 1999 but was eventually successful in buying the remaining interests. Lukoil Europe Holdings B.V.'s ownership share rose after two buyout offers from minority shareholders in 2004 to 93.25% by the beginning of 2005 (Vatansever, 2006, p. 23; Zashev, 2006, p. 119). Reports indicate that the company has consolidated its ownership to near 100% today.

Petrol AD was a monopolist that operated on the Bulgarian retail fuel market in the 1990s and had ownership of around one-third of all petrol stations in Bulgaria. In 1998–99, it was privatized by the Bulgarian Privatization Agency with three bidders expressing interest: a consortium of PAO Lukoil and Rosneft International Ltd.; Yukos Petroleum Bulgaria; and a consortium of Petrol Holding Group AD and OMV Bulgaria OOD. The latter two formed International Consortium Bulgaria and eventually won the tender for a 51% stake in Petrol AD. Since Petrol Holding Group AD already owned 24.2% of shares, International Consortium Bulgaria wound up with a 75.2% share in the company (Vatansever, 2006, pp. 24–28). Adnan Vatansever also adds the ownership changed in succeeding years as Petrol Holding AD (former Naftex Bulgaria Holding, formerly Yukos Petroleum Bulgaria) acquired the shares of Petrol Holding Group AD and of OMV Bulgaria OOD through an asset swap, and in 2004 owned 92.57% of Petrol AD (70.57% of Petrol AD's shares plus another 22% through its subsidiary Ros Oil EOOD) (Vatansever, 2006, p. 28).

That situation has recently changed again: in 2012, the Switzerland-based Credit Mediterranee (owned by Russian citizen Kirsan Ilyumzhinov) purchased 52.5% of Petrol Holding Group AD from the individuals behind the company, Denis Ershov (47.5%) and Alexander Melnik (5%) ('*Shrewd Move*', 2012; '*The President of*', 2012). Allegedly, Kirsan Ilyumzhinov is in negotiations with Mitko Sabev, who owns the remaining 47.5% of Petrol Holding Group AD, to buy the entire company. It is however highly improbable that Kirsan Ilyumzhinov is operating as a front for Moscow; his operational record instead presents the image of a multimillionaire businessman and acquisitive oligarch. During his presidency of the Republic of Kalmykia, he was repeatedly accused of financial irregularities and violations of federal law, but was never prosecuted. In November 2015, he was named as one of four Specially Designated Nationals providing support to the Government of Syria by the US Department of the Treasury for actions that included the facilitation of Syrian government oil purchases from ISIL (U.S. Department of the Treasury, 2015). US entities are generally prohibited from engaging in transactions with persons so designated.

The AMBO (Albanian Macedonian Bulgarian Oil) pipeline is a planned crude oil pipeline from Burgas, Bulgaria via FYROM to the Albanian maritime port of Vlorë, with a capacity of around 36 million tonnes per year.[6] Plans call for the pipeline to stretch 894.5 km, with four pumping stations along the way (three in Bulgaria, one in FYROM). Similar to the Burgas-Alexandroupolis project,[7] the primary logic of this pipeline is to bypass the Bosporus Strait and serve as an alternative route for Russian and Caspian oil not only to countries overland, but more importantly to the Atlantic market via the Vlorë oil terminal in Albania. The Balkan countries understand that the strategic importance of those countries that host any future pipeline from the Black Sea to the Adriatic will rise substantially, and there is accordingly stiff competition among them. Greece is obviously against the project and supports the Burgas-Alexandroupolis option; however, the AMBO pipeline actually avoids the Bosporus Strait entirely, terminating at Vlorë.

The project dates back to 1993, when it was presented for the first time. In November 1995, Vuko Tashkovich (a Macedonian immigrant to the USA) founded the Albanian Macedonian Bulgarian Oil Corporation; in January 1997, Edward Ferguson, Director of Oil & Gas Development

---

[6] Recalculation of the capacity of 750,000 barrels per day to millions of tonnes per year was based on the density and specific gravity of Brent crude (835 kg/m³). Using this method, a barrel of oil weighs 132.754393162 kg.

[7] See chapter on Greece.

in Louisiana-based Brown & Root, was appointed President and CEO of AMBO. In 2004, Bulgarian (Simeon Borisov Saxe-Coburg-Gotha), Macedonian (Vlado Bučkovski), and Albanian (Fatos Nano) prime ministers signed a political declaration, followed by a Memorandum of Understanding between country representatives and Edward Ferguson ('*AMBO Pipeline Deal*', 2007; '*AMBO Trans-Balkan*', 2004; Mendes, 2003). The AMBO project was supported by the US government: the feasibility study confirming its viability was sponsored by the USA (Lesser, Larrabee, Zanini, & Vlachos, 2001, p. 96). In 2006, the fourth Interstate Meeting was held in Tirana, where the draft of the trilateral convention on construction was initialled by delegation leaders from Albania, Bulgaria, and Macedonia (Stefanov, n.d.). The document was accepted and signed in 2007 at the trilateral convention in Skopje and later ratified by the parliaments of all three countries.

According to Gligor Tashkovich, the former Macedonian foreign investment minister and board member of AMBO, the project has kept a low profile in recent years while seeking major investors and awaiting the right timing for the project to materialize (Elliott, 2011). New life may be breathed into plans for the AMBO oil pipeline in the coming years, since Chevron Corporation finished a US \$5.4 billion expansion project in 2016 that increased the number of pumping stations along the pipeline from 5 to 15, almost doubling CPC capacity (see below) to 67 million tonnes annually. Tankers carrying the oil from Novorossiysk must inevitably sail through the Turkish Straits or else some other way to cross the Straits will have to be found. This could make the AMBO pipeline (or the Samsun-Ceyhan, see below) an attractive option.

The Caspian Pipeline Consortium (CPC) owns an oil pipeline whose starting point lies in Tengiz, Kazakhstan, and proceeds through Kazakhstan and Russia to Russia's Black Sea maritime port of Novorossiysk. Chevron Corporation holds a 15% stake in the CPC through the Chevron Caspian Pipeline Consortium Company and a 50% stake in Tengizchevroil, which operates the Tengiz and Korolev fields in Kazakhstan, the locations where the CPC pipeline starts. The company also has an 18% nonoperated working interest in the Karachaganak field. The Russian Federation has a presence in the CPC through OAO AK Transneft (31% in CPC; it manages the 24% stake of the Russian Federation and the 7% stake of the Russian CPC Company); LukArco B.V. (12.5% in CPC; it is a 100% subsidiary of PAO Lukoil); and Rosneft Shell Caspian Ventures Ltd. joint venture (7.5% in CPC; OAO NK Rosneft holds 51% of the joint venture and Royal

Dutch Shell plc 49%) (Chevron Corporation, 2015, pp. 23–24; OAO NK Rosneft, n.d.; Lukoil Overseas, n.d.; Caspian Pipeline Consortium, n.d.). Altogether the Russian share in CPC is 47.3%.

The AMBO pipeline has faced many obstacles, including the physical existence of adequate crude oil in the Caspian Region, intense competition with other infrastructural projects on the Balkan Peninsula, funding[8] for construction (which has not yet been secured because the project lacks a contractually guaranteed oil supply—a key prerequisite for a bank loan), and, of course, geopolitics in the region. The USA supported the project to improve its own position in the region and gain more leverage over exports from that area. Russia is interested in the project, too, because the pipeline would provide an alternative to oil exports through the Bosporus Strait controlled by Turkey. Obviously, US and Russian interests are at odds because Russia seeks to strengthen exports of Russian crude through the AMBO pipeline, unlike the USA, which is interested in Caspian oil reserves. And Turkey is in a similar position as it is with the Burgas-Alexandroupolis project: on the one hand, the pipeline would represent a threat to Turkey's position as the sole monopolist controlling the transit of energy resources in the region; on the other, it would solve the crucial problem of constantly rising tanker traffic through the Bosporus.

Since Bulgarian Prime Minister Boyko Borisov led his first (2009–13) and second (2014–) centre-right governments, Bulgarian and Russian relations have deteriorated because of Bulgaria's abandonment of the Burgas-Alexandroupolis oil pipeline project; its permission for three US military bases on its territory; the selection of Westinghouse Electric Company LLC's AP1000 design for Unit 7 of the Kozloduy Nuclear Power Plant in mid-2013 over Rosatom State Nuclear Energy Corporation's VVER design; and its support for Western sanctions against Russia. Bulgaria also stopped construction on the South Stream gas pipeline project due to an infringement procedure against the country for non-compliance with European rules on energy competition public procurement procedures (*'Bulgaria's Government'*, 2014). The project was eventually halted by Vladimir Putin in December 2014, and shortly afterwards exchanged for the Turkish Stream (also referred to as 'TurkStream'). Putin primarily blamed Bulgaria for the failure of the project, but in reality

---

[8] Among those that have expressed interest in the project are Overseas Private Investment Corporation (US Development Agency), Eximbank, Credit Suisse First Boston, European Union, European Bank for Reconstruction and Development, and others.

he was unhappy about EU energy competition laws that infringed upon Russian interests in the South Stream. Although the Turkish Stream project was shelved after the downing of a Russian fighter jet by Turkey, it was revived after a year once Russo-Turkish relations normalized. The Turkish Stream project is currently progressing according to plan with first deliveries planned to take place in late 2019–early 2020. As for Bulgaria, Borisov government's goal seems to be establishing Bulgaria as a gas hub and ultimately also reach improvements in relations with Russia. At the same time, Bulgaria still hopes that some supplies from Turkish Stream may eventually also flow to Bulgaria as Borisov's government is focused on dealing with the country's dependency on Russian Federation carbohydrates rather than supporting projects in the Russian interest.

Russian Federation, however, is aware of the geopolitical importance of Bulgaria for any Black Sea infrastructural project and reconciliation came shortly after, in August 2015. Even Borisov made an effort to normalize relations by making the impression that Bulgaria was not supporting Russian sanctions out of conviction but because of outside pressure (Boyadjiev & Andreev, 2015). This geopolitical development is likely connected to expansion of the Caspian Pipeline Consortium.

6.3   Research Indicator Assessment

6.3.1   Active Support by Russian State Representatives for Energy Enterprises and Their Activities Abroad

This indicator was positive several times in the past, especially during privatization of the Neftochim refinery. The Russian government did promote Russian involvement strongly, but this support was hardly coordinated (Vatansever, 2006, p. 30). Two Russian companies, PAO Lukoil and Yukos Petroleum Bulgaria, competed against each other, and the Russian government's support shifted over a brief period from one company to the other.

Russian support for the Burgas-Alexandroupolis oil pipeline project was also strong, but after years of hesitation by the Bulgarians, what had been clear support morphed into the wish for a clear answer. If Bulgaria decided against the Burgas-Alexandroupolis project, the Russian Federation was prepared to deepen its negotiations with Turkey over the Samsun-Ceyhan oil pipeline (Trans-Anatolian Pipeline) (Geropoulos, 2009). The Samsun-Ceyhan is another option for bypassing the Bosporus

Strait as an alternative route for Russian and Caspian oil. The planned pipeline, with a capacity of around 48 million tonnes annually, would start at Ünye near Samsun in Northern Turkey (on the shores of the Black Sea) and lead to the Southern city of Ceyhan on the Mediterranean coast, thus relieving traffic in the Bosporus and Dardanelles Straits. Bulgaria eventually decided to withdraw from the Burgas-Alexandroupolis project, but Russian Energy Minister Alexander Novak was quoted in 2013 as saying sending oil through the Straits would be as much as 40% cheaper (*'Turkey's Samsun-Ceyhan'*, 2013). In view of its lack of competitiveness, the project was shelved even before Russia-Turkey relations deteriorated sharply in the aftermath of Turkey's shooting down a Russian fighter jet.

6.3.2   As a Foreign Supplier, Russia Rewards Certain Behaviours and Links Energy Deals to the Client State's Foreign Policy Orientation

As with Greece, some used to call Bulgaria the Trojan Horse of Russia in the EU (Williams & Tsolova, 2014). Such a characterization might be put down to the memory of the intimate relations between Sofia and Moscow during the communist era or the recognition that Bulgaria is highly dependent on Russian energy (over 90% of natural gas, over 80% of crude oil, and full dependency on nuclear fuel), but other factors include Russian tourism, real property ownership, and bilateral trade connections (more than 68,000 Russian tourists visited the country last year, more than 300,000 Russians own real estate in Bulgaria, and Russian companies have invested more than US $2 billion in Bulgaria; Davydenko, 2014). As stated above, Russian-Bulgarian relations mutate considerably depending upon who is in government in Bulgaria. The centre-right governments of Boyko Borisov have been far more cautious about Russian involvement in Bulgaria, though currently Bulgaria and Russia are undergoing a notable rapprochement after Putin's accusation that Bulgaria was the sole reason the Burgas-Alexandroupolis and South Stream projects did not materialize. This might be connected to current developments in the Caspian Pipeline Consortium (see above).

A strict connection of this indicator to the Bulgarian crude oil sector was not found. Some argue the Russian reaction to Sofia's decision not to continue with Burgas-Alexandroupolis and South Stream projects was to block Bulgaria's involvement in any alternative Balkan pipeline project and Russia's turn to Turkey instead (*'Bulgaria Ousted'*, 2011).

6.3.3   Abuse of Infrastructure (e.g. Pipelines) and Differential Pricing to Exert Pressure on the Client State

As regards crude oil infrastructure, crude is imported using tanker ships. Aside from the short 15 km pipeline from the Burgas oil terminal Rosenets to the Neftochim refinery, no crude oil pipelines are used. The pipeline and the refinery are owned by Lukoil Neftochim Burgas AD, a company owned by Lukoil Europe Holdings B.V. (a subsidiary of the Russian firm PAO Lukoil). Oil is imported mainly from the Russian Federation (over 80%), which makes Russia the key supplier of crude oil at approximately 3 million tonnes annually. The refining sector as well as distribution and retail are in private hands with no effective oversight by the Bulgarian government. The very high concentration of market power in the crude oil sector led to an abuse of the company's position on the wholesale market. In 2012, the Commission for the Protection of Competition probed Lukoil Bulgaria Ltd. (a subsidiary of PAO Lukoil) for abusing its monopoly power over the retail market for diesel and A95 unleaded gasoline. The Commission found there had been an illegal agreement between Lukoil Bulgaria Ltd., Rompetrol Bulgaria EAD, Naftex Petrol EOOD, and OMV Bulgaria OOD to coordinate prices (Center for the Study of Democracy, 2014, p. 66). However, this development is quite difficult to connect with pressure on the client state, as it instead represents a joint effort by multiple retail companies to maximize capitalization. Also, the Commission itself later said that Lukoil Bulgaria Ltd.'s market behaviour did not constitute a breach of competition law.

6.3.4   Efforts to Take Control of the Energy Resources, Transit Routes, and Distribution Networks of the Client State

After Lukoil Europe Holdings B.V. won the privatization process for the Neftochim refinery, PAO Lukoil company fundamentally strengthened its position in the country's petroleum sector to include full control of imports, full ownership of the midstream (refinery) and of wholesale (through LITASCO SA), and a strong position in retail through its Lukoil Bulgaria Ltd. petrol stations network.

The Bulgarian Petroleum and Gas Association was founded in 1999. Among its main objectives are to contribute to the process of demonopolization of the petroleum industry, develop free initiative and the basic principles of the market economy, prevent the misuse of a monopolist market position, and forestall unfair competition and other actions that might lead to a violation of market principles and negatively impact the market for petroleum and gas products (Bulgarian Petroleum and Gas Association, n.d.). All market actors in the Bulgarian petroleum market are members of the association.

Since that time, no strategic behaviour aimed at controlling infrastructure has been found (mainly because the majority is already under Russian control). However, NIS Petrol EOOD (a subsidiary of NIS Gazprom Neft), with a network of 35 Gazprom petrol stations in Bulgaria, has entered the retail market recently with a plan to operate a network of 250 units under the Gazprom brand in Serbia, Bulgaria, Romania, and Bosnia by the end of 2015, with the strategic goal of becoming a leader in the Balkan retail car fuel market, according to Anatoly Cherner, Deputy CEO for Logistics, Processing and Sales, Gazprom Neft ('Gazprom Petrol', 2013; Daskalovic, 2015).

6.3.5   Disruption (by Various Means) of Alternative Supply Routes/Sources of Supply

Given the technology used at the Neftochim refinery, long-term imports of Russian crude are essentially a certainty. The smaller exporters to Bulgaria, such as Kazakhstan and Iran (offering a similar type of crude oil, heavy and sulphurous), use the CPC pipeline. The evidence concerning Russian behaviour in the Black Sea and Caspian regions shows Russian efforts to control exports through the CPC. But these efforts are difficult to connect to Russian-Bulgarian relations only—Caspian resources and pipeline policy are part of a much bigger geopolitical arena.

6.3.6   Efforts to Gain a Dominant Market Position in the Client Country

The Sofia-based Center for the Study of Democracy stated in its 2014 report that although the Commission for the Protection of Competition found in 2012 that Lukoil Bulgaria Ltd.'s market behaviour does not constitute a breach of competition law, there have been continuous allegations by non-governmental organizations and large fuel clients that by providing wholesale buyers with discounted fuel prices in exchange for loyalty, the company dominates the downstream market, in effect preventing the entry of international competition (Center for the Study of

Democracy, 2014, p. 66). Also, the entry into the Bulgarian market of NIS Petrol EOOD and NIS Gazprom Neft's strategic plans in the Balkan Peninsula should be approached with caution.

6.3.7   Efforts to Eliminate Competitive Suppliers

Due to the technology used at the Neftochim refinery (designed primarily to refine Russian Export Blend) competitive suppliers are automatically limited to those who offer the same crude oil type (heavy and sulphurous), such as Kazakhstan, Iran, and Iraq.

The abuse of the wholesale market position of Lukoil Bulgaria Ltd. through an agreement with three other companies to coordinate price could be taken as an effort to eliminate competition by partitioning the market.

6.3.8   Acting Against Liberalization

The abuse of the wholesale market position of Lukoil Bulgaria Ltd. through the agreement with three other companies to coordinate price could also be taken as an activity that flies in the face of liberalization of the petroleum sector. Preventing the entry of international competition through domination of the wholesale market and discounted fuel prices in exchange for loyalty, as reported by the Center for the Study of Democracy (2014, p. 66), is compelling evidence that Lukoil is acting against liberalization and market principles. Also, during every tax crisis involving Lukoil, the company has threatened to cut the supply of finished fuel products to the market, and hence has been able to influence milder subsequent treatment by the government (Center for the Study of Democracy, 2014, p. 67).

6.3.9   Diminishing the Importance and Influence of Multilateral Regimes Such as the EU

Indications were noted in connection to the Burgas-Alexandroupolis pipeline and the AMBO pipeline projects in the 1990s and 2000s. The Russian Federation preferred to negotiate with the individual countries concerned. Negotiations with Bulgaria (Greece, Macedonia, and Albania) have taken place on a bilateral and/or trilateral basis.

6.3.10   Attempts to Control the Entire Supply Chain (Regardless of Commercial Rationale)

Russia's Lukoil Europe Holdings B.V. controls crude imports to the Neftochim refinery, owns the Neftochim refinery itself (though it took several years to buy out the stakes of all minority shareholders), the Burgas oil terminal Rosenets, and the 15 km pipeline from Rosenets to the Neftochim refinery. The company enjoys a dominant position on the wholesale market through LITASCO SA (PAO Lukoil's 100% subsidiary) and a strong position on the retail market, as well, through its Lukoil Bulgaria Ltd. petrol station network. There is no evidence that the acquisition and developmental activities of Lukoil Europe Holdings B.V. lacks a commercial or market rationale or that it is only minimal.

6.3.11   Economically Irrational Steps Taken to Maintain a Particular Position in the Client State's Market

No such steps were found to have been taken in Bulgaria. It is difficult to see the operation of Lukoil Europe Holdings B.V. or NIS Gazprom Neft in Bulgaria not being based upon an economic rationale (Table 6.3).

Table 6.3   Summary of indicators

Indicator	Found	Found with
Active support by Russian state representatives for energy enterprises and their activities abroad	Yes	PAO Lukoil, Yukos Petroleum Bulgaria, Pipeline Consortium Burgas-Alexandroupolis Ltd. (OAO AK Transneft, 33.34%; OAO NK Rosneft, 33.33%; PAO Gazprom Neft, 33.33%)
As a foreign supplier, Russia rewards certain behaviours and links energy deals to the client state's foreign policy orientation	No	–
Abuse of infrastructure (e.g. pipelines) and differential pricing to exert pressure on the client state	Inconclusive	–
Efforts to take control of the energy resources, transit routes, and distribution networks of the client state	Inconclusive	–

(continued)

Table 6.3   (continued)

Indicator	Found	Found with
Disruption (by various means) of alternative supply routes/sources of supply	Inconclusive	–
Efforts to gain a dominant market position in the client country	Yes	Lukoil Bulgaria Ltd.
Efforts to eliminate competitive suppliers	Yes	Lukoil Bulgaria Ltd.
Acting against liberalization	Yes	Lukoil Bulgaria Ltd.
Diminishing the importance and influence of multilateral regimes such as the EU	Yes	–
Attempts to control the entire supply chain (regardless of commercial rationale)	No	–
Economically irrational steps taken to maintain a particular position in the client state's market	No	–

Source: Author

Note: *Inconclusive* means some indications of this behaviour were found, but not of a shape, size, or importance to be ascribed to strategic behaviour and thus fulfil the indicator

Sources

AMBO Pipeline Deal Clears Another Hurdle. (2007, April 2). *Pipelines International.* Retrieved from http://pipelinesinternational.com/news/ambo_pipeline_deal_clears_another_hurdle/011027/

AMBO Trans-Balkan Pipeline Agreement Finally Signed. (2004, December 29). *Balkanalysis.com.* Retrieved from http://www.balkanalysis.com/blog/2004/12/29/ambo-trans-balkan-pipeline-agreement-finally-signed/

Angelov, I. (2012, June 20). Russian Oligarch Buys Bulgarian 'Petrol Holding'. *NewEurope.eu.* Retrieved from http://neurope.eu/article/russian-oligarch-buys-bulgarian-petrol-holding/

Boyadjiev Y., & Andreev, A. (2015, August 21). Why Putin Is Courting Bulgaria. *Deutsche Welle.* Retrieved from http://www.dw.com/en/why-putin-is-courting-bulgaria/a-18663257

BP plc. (2015). *BP Statistical Review of World Energy June 2015.* London: BP plc. Retrieved from http://www.bp.com/content/dam/bp/en/corporate/pdf/bp-statistical-review-of-world-energy-2015-full-report.pdf

Bulgaria Ousted from Russia's South Stream Pipe. (2011, August 31). *EurActiv. com*. Retrieved from http://www.euractiv.com/section/europe-s-east/news/ bulgaria-ousted-from-russia-s-south-stream-pipe/

Bulgaria's Government to Collapse over South Stream. (2014, June 10). *EurActiv. com*. Retrieved from http://www.euractiv.com/section/energy/news/bulgaria-s-government-to-collapse-over-south-stream/

Bulgarian Petroleum and Gas Association. (n.d.). Retrieved from http://bpga.net/

Bulgarian Petroleum Refinery Ltd. (n.d.). Retrieved from http://www.bpr-bg.com/

Caspian Pipeline Consortium. (n.d.). Retrieved from http://www.cpc.ru/

Center for the Study of Democracy. (2014). *Energy Sector Governance and Energy (IN)Security in Bulgaria*. Sofia: Center for the Study of Democracy.

Chevron Corporation. (2015). *Supplement to the Annual Report*. San Ramon: Chevron Corporation. Retrieved from http://www.chevron.com/documents/pdf/annual-report-supplement-2015.pdf

Chimimport Plc. (2013). *Presentation of 'Chimimport' AD*. Retrieved from http://www.chimimport.bg/en/

Chimimport Plc. (2014). *Annual Activity Report 2014*. Retrieved from http://www.chimimport.bg/uf//dl/statements/individual.eng/Annual%20Report%202014%206C4%20ENG.pdf

Daskalovic, D. (2015, May 6). NIS Raises Count of Gazprom Fuel Stations in Serbia to 13. *SeeNews*. Retrieved from http://wire.seenews.com/news/nis-raises-count-of-gazprom-fuel-stations-in-serbia-to-13-475310

Davydenko, A. (2014, August 8). *Bulgaria-Russia: The Past, the Present, and the Future*. On the 135th Anniversary of the Establishment of Diplomatic Relations Between the Countries. *International Affairs*. Retrieved from http://en.inter-affairs.ru/experts/542-bulgaria-russia-the-past-the-present-and-the-future-on-the-135th-anniversary-of-the-establishment-of-diplomatic-relations-between-the-countries.html

Elliott, S. (2011, November 11). Interview: Trans-Balkan AMBO Oil Pipeline Still Viable Project. *Platts McGraw Hill Financial*. Retrieved from http://www.platts.com/latest-news/oil/london/interview-trans-balkan-ambo-oil-pipeline-still-8572004

Enerdata. (2013). *Bulgaria Energy Report*. London: Enerdata.

Gazprom Petrol Stations Network Opened in Bosnia and Herzegovina. (2013, July 24) *Gazprom Neft Press Release*. Retrieved from http://www.gazprom-neft.com/press-center/news/1095253/

Geropoulos, K. (2009, September 7). Borisov's Blind Man's Bluff with Putin. *Novinite.com*. Retrieved from http://www.novinite.com/articles/107574/ Borisov's+Blind+Man's+Bluff+with+Putin

International Energy Agency. (n.d.). Retrieved from http://www.iea.org/

Ivanova, I., & Tsolova, T. (2011, July 27). UPDATE 3-Bulgaria Forces LUKOIL Neftochim Refinery Shutdown. *Reuters.com*. Retrieved from http://www. reuters.com/article/bulgaria-lukoil-idUSL6E7IR0BX20110727

Jirušek, M., Vlček, T., & Henderson, J. (2017). Russia's Energy Relations in Southeastern Europe: An Analysis of Motives in Bulgaria and Greece. *Post-Soviet Affairs, 33*(5), 335–355. https://doi.org/10.1080/10605 86X.2017.1341256

Lesser, I. O., Larrabee, F. S., Zanini, M., & Vlachos, K. (2001). *Greece's New Geopolitics.* Santa Monica, Arlington, and Pittsburgh: RAND.

LITASCO SA. (n.d.). Retrieved from http://www.litasco.com/

Lukoil Neftochim Burgas AD. (n.d.). Retrieved from http://www.neftochim.bg/

ЛУКОЙЛ Нефтохим Бургас АД. (2014). *Финансов отчет за годината приключваща на 31 декември 2014.* Бургас: ЛУКОЙЛ Нефтохим Бургас АД. Retrieved from http://www.neftochim.bg/assets/components/lukoil/pdf/bg/financial_reports/2014.pdf

Lukoil Overseas. (n.d.). Retrieved from http://lukoil-overseas.com/

Mendes, A. J. (2003, July 29). Balkans Crisis Supports US Corporate Interests. *Centre for Research on Globalisation.* Retrieved from http://www.globalresearch.ca/articles/MEN307A.html

Ministry of Economy of the Republic of Bulgaria. (2012). *Bulletin on the State and Development of the Energy Sector in the Republic of Bulgaria.* Sofia: Ministry of Economy of the Republic of Bulgaria. Retrieved from http://www.mi.government.bg/files/useruploads/files/budget/bulletin_energy_2012_eng.pdf

Ministry of Environment and Water of the Republic of Bulgaria. (2014). *National Inventory Report 2014 for Greenhouse Gas Emissions. Submission Under the UNFCCC and the Kyoto Protocol.* Sofia: Executive Environment Agency at the Ministry of Environment and Water.

Nitzov, B., Stefanov, R., Nikolova, V., & Hristov, D. (2010, April 6). *The Energy Sector of Bulgaria.* Atlantic Council Issue Brief. Retrieved from www.atlanticcouncil.org/images/files/publication_pdfs/403/BulgariaEnergy_ECIssueBrief.pdf

OAO NK Rosneft. (n.d.). Retrieved from http://www.rosneft.com/

Shrewd Move: President of FIDE Buys into Bulgarian Fuel Major Petrol. (2012, June 20). *RT.com.* Retrieved from https://www.rt.com/business/bulgaria-petrol-ilyumzhinov-deal-287/

Stankova, R. (n.d.). *Bulgaria, Organization of Executing the Obligations Following the Membership of the Republic of Bulgaria in the European Union in the Sphere of Compulsory Oil Stocks.* Presentation of Deputy Head Director of the State Agency 'State Reserve and War-time Stocks'. Retrieved from https://www.energy-community.org/portal/page/portal/ENC_HOME/DOCS/622178/0633975AAD5C7B9CE053C92FA8C06338.PDF

Stefanov, S. (n.d.). *Oil Pipeline Projects to Bypass the Turkish Straits for Oil Transportation.* Ministry of Economy and Energy Bulgaria. Retrieved from http://www.energycharter.org/fileadmin/DocumentsMedia/Presentations/CBP-Straitsbypass.pdf

The President of the World Chess Federation and Ex-president of Republic of Kalmykia Kirsan Ilyumzhinov Has Bought a 52.5% Stake in Bulgaria's Largest Fuel Company Petrol Holding. (2012, June 21). *Kalmykia.eu*. Retrieved from http://www.kalmykia.eu/2012/the-president-of-the-world-chess-federation-and-ex-president-of-republic-of-kalmykia-kirsan-ilyumzhinov-has-bought-a-52-5-stake-in-bulgaria%E2%80%99s-largest-fuel-company-petrol-holding/

Turkey's Samsun-Ceyhan Oil Pipeline Shelved. (2013, April 23). *United Press International*. Retrieved from http://www.upi.com/Business_News/Energy-Industry/2013/04/23/Turkeys-Samsun-Ceyhan-oil-pipeline-shelved/24281366711120/

U.S. Department of the Treasury. (2015, November 25). *Treasury Sanctions Networks Providing Support to the Government of Syria, Including For Facilitating Syrian Government Oil Purchases from ISIL*. Washington, DC: U.S. Department of the Treasury. Retrieved from https://www.treasury.gov/press-center/press-releases/Pages/jl0287.aspx

Vatansever, A. (2006). *Russian Involvement in Eastern Europe's Petroleum Industry, The Case of Bulgaria*. London: GMB Publishing Ltd.

Williams, M., & Tsolova, T. (2014, April 10). Bulgaria Torn Between Old Friends and New Partners over Crimea. *Reuters*. Retrieved from http://www.reuters.com/article/us-bulgaria-russia-idUSBREA390AW20140410

Zashev, P. (2006). Russian Companies in Forthcoming EU Member States: A Case of Lukoil in Bulgaria. In K. Liuhto (Ed.), *Expansion or Exodus: Why Do Russian Corporations Invest Abroad?* (pp. 109–128). Binghamton: The International Business Press.

CHAPTER 7

Republic of Croatia

7.1 Crude Oil Sector General Information

7.1.1 Introduction and Upstream

Croatia is a Northwestern Balkan country with around 4.2 million inhabitants spread over nearly 57,000 km². The country borders the Adriatic Sea on the West and is neighbour to Slovenia, Hungary, Serbia, Bosnia and Herzegovina, and Montenegro.

Croatia's crude oil needs range between 4 and 5 million tonnes annually, with 15–20% of this amount covered by domestic production. The rest is imported through the JANAF (Jadranski Naftovod) pipeline system. Russian export blend (REB) used to be the major oil type refined in Croatia, but INA—Industrija nafte d.d., owner of the country's refineries, has focused on broadening the crude oil portfolio. It eventually moved in the direction of increasing the share of alternative crude types to 74% of all imported processed crude grades in 2015 (INA—Industrija nafte d.d., 2015, p. 47). Russian export blend is thus a minority oil type, refined in Sisak, and accounted for only 0.58 mta of 2015 imports (Table 7.1).

In 2014, crude oil constituted 37.4% of the total primary energy supply, petroleum products 39.5% of total final consumption (International Energy Agency, n.d.). The majority of these products (69.5%) were consumed in the transport sector.

In 2013, Croatia introduced a modern oil and gas licencing process as part of an effort to stimulate the Croatian economy. Official estimates say

Table 7.1   Republic of Croatia crude oil data

	2008	2009	2012	2014
Consumption	4614	5016	4065	3125
Production	818	786	612	603
Import dependency	82.3%	84.3%	84.9%	80.7%

Source: INA—Industrija nafte d.d., 2009, p. 9, 2014, p. 4

Note: Figures in thousands of tonnes; percentage calculations by T. Vlček

the state budget could reach around EUR 1 billion annually taking into account the indirect effects of industrial development and employment (Veselica, 2015). There is the potential risk of a negative impact on the coastline, keeping in mind that tourism accounts for about 20% of Croatia's gross domestic product (GDP). In April 2014, the first offshore licence round was announced, with 29 exploration blocks in Croatian waters in the Adriatic Sea. Licences for exploration and exploitation were then awarded for ten blocks to US-based Consortium Marathon Oil Corporation and OMV AG (seven blocks), Consortium Eni S.p.A. and Medoilgas Italia S.p.A. (one block), and INA—Industrija nafte d.d. (two blocks). The first onshore licencing round was announced in February 2015 with six exploration blocks. All six were awarded to three companies: Vermilion Zagreb Exploration d.o.o. (four blocks, a subsidiary of Canada's Vermilion Energy Inc.), Nigeria's Oando PLC (one block), and INA—Industrija nafte d.d. (one block) (Agencija za ugljikovodike, n.d.).

AZU (Agencija za ugljikovodike, the Croatian Hydrocarbon Agency) was created to oversee the preparation and fulfilment of the exploration and exploitation contracts and Production Sharing Agreements (PSAs). But signing the Production Sharing Agreements was postponed so that the procedures could be implemented before the signing took place. It was not until June 2016 that the first PSAs were signed with INA—Industrija nafte d.d. (one block) and Vermilion Zagreb Exploration d.o.o. (four blocks)—and still only for the onshore blocks. The PSAs for the offshore blocks came under sharp criticism from the public and because of the harm they might do to the tourism industry. In the meantime, Consortium Marathon Oil Corporation and OMV AG withdrew from the signature process, since three of their blocks were located in a disputed area that lay between Croatia and Montenegro ('*Marathon Oil and*', 2015).

In the end no offshore PSAs were signed, and none are slated for signature any time soon. The new licencing process was developed under the centre-left government of Prime Minister Zoran Milanović, but the 2016 centre-right government of Tihomir Orešković (replaced again in October 2016 by Andrej Plenković) halted the process precisely because of the harm it might do to the tourism industry. Tomislav Panenić, Minister of the Economy, confirmed that plans for oil and gas exploration in the Adriatic had been abandoned ('*Contracts Signed*', 2016; '*Croatia Signs*', 2016).

As a result, over a 60-year period, all oil exploration and production in Croatia has been carried out by a single company with 35 oil fields, the historical national champion INA—Industrija nafte d.d., owned today by Hungary's MOL Rt (49.08%), the Republic of Croatia (44.84%), and other institutional and private investors (6.08%) (INA—Industrija nafte d.d., 2015, p. 20). The company also has concessions in Angola, Egypt, and Syria (temporarily suspended since 2012).

7.1.2   Midstream

Crude oil is imported to Croatia's two refineries in Urinj (Rijeka) and Sisak from Omišalj, a port on the Croatian Island Krk. The stretch of the pipeline that runs through Croatia is part of a system called Jadranski Naftovod (JANAF).[1] More than ten entities are included in its ownership structure, among them the Government of Croatia, with a 78.509% share[2] (*Jadranski Naftovod d.d.*, n.d.). JANAF also connects with the Adria pipeline, which runs between the town of Gola on the border between Croatia and Hungary, and the Hungarian refinery Duna, located near Budapest in Százhalombatta.

The JANAF system was constructed in 1974–79 to open a new route that would allow oil to be obtained from alternate suppliers at a cheaper price, and also to enable a stronger negotiating basis with Russia via diversification. The Balkan states found new suppliers among the Arabic countries and, according to informed sources, got some of that oil in exchange for arms. The pipeline does not, however, fulfil its original purpose. First constructed to bring in oil from the Middle East, it currently serves to

---

[1] Historically also called JUNO (Jugoslavski Naftovod).

[2] By means of the Agency for State Property Management, Croatian Agency for Supervision of Pension Funds and Insurance, State Agency for Insuring Deposits, and Bank Rehabilitation and Restructuring and Sale Centre.

transport oil from Russia to the refineries in Sisak, Croatia (IEA, 2011, p. 51), and Százhalombatta, Hungary, using the reverse direction of flow. The Bosnian refinery in Brod and the Serbian refineries in Novi Sad and Pančevo are supplied via the JANAF system from Omišalj to Serbia.[3]

The JANAF consists of 622 km of pipeline with a maximum transport capacity of 20 mta on the Omišalj-Sisak and Sisak-Gola sections. In 2013, 5.4 million total tonnes of crude oil were transported through the system (Croatian Energy Regulatory Agency, 2014, p. 134). The pipeline transport capacity is 20 mta (Omišalj-Sisak), 9 mta (Sisak-Novi Sad and Sisak-Brod), 6 mta (Novi Sad-Pančevo), 20 mta (Sisak-Gola), 14 mta (Gola-Százhalombatta), and 0.6 mta (Virje-Lendava). Domestic crude oil and condensate from oil fields in Moslavina is transported via the local oil pipeline and from oil fields in Slavonia via the Sava River. Croatia has three maritime ports capable of receiving oil and oil products: Omišalj, Zadar, and Ploče.

7.1.3   Downstream

There are two crude oil refineries in Croatia, one in Urinj (Rijeka) and one in Sisak. Both are owned and operated by INA—Industrija nafte d.d. Refining capacity is 4.5 mta in Urinj (Rijeka) and 2.2 mta in Sisak. In addition to these two crude oil refineries, INA—Industrija nafte d.d. operates two lubricant production plants, in Rijeka (with a capacity of 450,000 tonnes annually) and Zagreb (45,000 tonnes annually), through its subsidiaries INA MAZIVA Ltd. and Maziva Zagreb d.o.o. (Table 7.2).

Table 7.2 Capacity of Croatian refineries as of 2016

Refinery	Owner	Refining capacity	Nelson complexity index (2016)	Year established
Sisak	INA—Industrija nafte d.d.	2.2 mta	6.1	1956
Urinj (Rijeka)	INA—Industrija nafte d.d.	4.5 mta	9.1	1965
Rijeka[a]	INA MAZIVA Ltd.	0.45	–	1883
Zagreb[a]	Maziva Zagreb d.o.o.[b]	0.045	–	1927

Source: Compiled by T. Vlček from public sources

[a]Lubricant production plant
[b]100% owned by INA MAZIVA Ltd.

---

[3] More on the JANAF and Adria systems in Vlček (2015, pp. 125–136).

The Sisak refinery is flexible in terms of the types of crude it processes. Crude oil and condensate are transported from domestic oil fields in Moslavina via the local oil pipeline and from domestic oil fields in Slavonia by barge on the Sava River. REB and Azeri crude is transported through the JANAF pipeline (MOL Group, n.d.). Production focuses mainly on diesel and gasoline production.

The second refinery is known alternatively as Urinj (after the Urinj Peninsula), Kostrena, Bakar (after the actual coastal cities/areas where the refinery stands), or Rijeka (after the largest nearby town). The Urinj refinery also flexibly processes different types of crude (from Russia, the Middle East, Africa, the Caspian Basin, and other sources). These crudes are imported through the port of Omišalj and subsequently through the 7.2 km Omišalj-Rijeka submarine pipeline section. Production focuses mainly on diesel, gasoline, and liquefied hydrogen production (MOL Group, n.d.).

INA—Industrija nafte d.d. invests in both refineries to maintain their competitiveness; the focus is on enhancing the production of lighter, higher-value derivatives. A major upgrade programme was started in the Urinj refinery in 2015 and in the Sisak refinery in 2008. But because of the global oil situation, the Sisak refinery is at risk of being closed by a decision of the parent company, MOL Rt, and used as a distribution and logistics centre (Pavlic, 2016).

There are around 815 petrol stations in Croatia, 388 (47.6% of the market) of which are operated by INA—Industrija nafte d.d. (INA—Industrija nafte d.d., 2015, p. 5). Petrol Hrvatska d.o.o. (a 100% subsidiary of Slovenian Petrol d.d., Ljubljana) operates a network of 102 petrol stations (12.5% of the market); Crodux derivati dva d.o.o., a Croatian company that took over the petrol stations of OMV Hrvatska d.o.o. in 2013, operates 66 petrol stations (8.1% of the market). LUKOIL Croatia d.o.o. runs a network of 52 petrol stations (6.4% of the market), and TIFON d.o.o. (a 100% subsidiary of MOL Rt) operates 41 petrol stations (5% of the market). These five entities control more than 80% of the Croatian retail market. Besides these companies, around 40 other smaller enterprises operate small-scale petrol station networks.

7.2   RECENT MARKET DEVELOPMENTS
AND RUSSIAN ACTIVITIES

Five Russian companies have shown interest in starting or developing business in Croatia. These are PAO Lukoil, PJSC Gazprom Neft, PAO NK Rosneft, OAO Zarubezhneft, and PAO Transneft.

PAO Lukoil is already present in the country with a network of 52 petrol stations owned by its subsidiary LUKOIL Croatia d.o.o. It entered the Croatian market in 2008, when Lukoil Europe Holdings (a 100% subsidiary of PAO Lukoil) purchased the nine-station network of Europa Mil d.o.o. (with five plots of land on which to build filling stations and the oil product hub TERMINAL DUNAV d.o.o. on the River Danube). This purchase of a small unimportant company was part of a planned strategy to become the second-largest player in Croatia by 2011, with 100–150 petrol stations, and to compete against Croatia's INA—Industrija nafte d.d. (*'Lukoil to Buy'*, 2008; *'Lukoil Wants'*, 2008). The company also negotiated to purchase TIFON d.o.o. but was ultimately successful, allegedly because MOL Rt had made a pre-agreement to take over of TIFON d.o.o. with its owner Ivan Čermak (*'Russians to Enter'*, 2007). In 2010, the emergent company, LUKOIL Croatia d.o.o., purchased Crobenz d.d., with 14 petrol stations (a 100% subsidiary of INA—Industrija nafte d.d.). As of 2016, the company still operated only 52 petrol stations; it did not meet its strategic plans. Besides the obvious reason behind these purchases—to sell its products from refineries in Bulgaria and Romania—another reason might be the wish to develop its position in Croatia before the country's entrance to the European Union so that it might later benefit from access to the EU market.

PJSC Gazprom Neft is another company that has shown an interest in entering the Croatian market, specifically in oil and gas production in the Pannonian Basin and on the Adriatic Sea shelf and created Gazprom Neft Adria d.o.o. in 2012 to engage in the new licencing process Croatia introduced in 2013. Gazprom Neft Adria d.o.o. was not awarded the licence, however, due to its failure to submit the appropriate bank guarantees[4] (*'Gazprom Neft Without'*, 2015). PJSC Gazprom Neft tried to purchase OMV Hrvatska d.o.o. in 2013, competing unsuccessfully along with Crodux derivati dva d.o.o. PJSC Gazprom Neft then tried to negotiate the purchase of Crodux derivati dva d.o.o. and its 66 petrol stations with owner Ivan Čermak. The negotiations are reported to have been mediated by former Croatian President Stjepan Mesić, who is known to have good business connections in Russia. But PJSC Gazprom Neft did not succeed in this transaction either (*'Crodux Could'*, 2016; *'GazpromNeft to Buy'*, 2013; *'Vlasnik Croduxa'*, 2016).

---

[4] There are allegations that the USA, which had previously opposed the entry of the Russians in INA—Industrija nafte d.d., may have influenced the process.

Since 2001, negotiations have been under way on using the Adria and JANAF pipeline systems to export Russian oil through the Omišalj terminal.[5] In December 2002, Croatia and Russia concluded an agreement to support a project integrating the Adria and Druzhba Pipelines by reversing the flow in the Sisak-Omišalj section of the JANAF system.[6] This would precisely invert the original intent of transporting oil inland as an interstage between Russian export pipelines and tanker transport from Omišalj (Socor, 2010). PAO NK Rosneft stated that their main interest was JANAF's Omišalj terminal for petroleum export on the Adriatic Sea and its infrastructure, by which PAO NK Rosneft could deliver oil to the USA. The Russian Federation therefore made efforts to acquire a stake in INA—Industrija nafte d.d. (11.8% owner of Jadranski Naftovod d.d.). The Croatian company was opened for privatization (49% of shares) and PAO NK Rosneft competed in 2003 with the Hungarian MOL Rt and Austria's OMV AG. PAO NK Rosneft withdrew at the last moment because Croatia rejected sale of the controlling stake. The Croatian government named MOL Rt the winner, which thus strengthened the security of supply by blocking plans for reverse flow operations on the JANAF-Adria pipeline systems (Orban, 2008, p. 100).

The main interest of the Russian Federation in Croatia therefore was and probably still is to acquire the terminal in Omišalj and control the route to it. In the 2000s, PAO NK Rosneft negotiated direct Russian supply to the USA with Marathon Oil Corporation ('*Russians to Enter*', 2007). The port of Omišalj and INA—Industrija nafte d.d.'s refining capacities were part of the strategy for exporting oil and oil products overseas.

Informed sources say the idea of transporting Russian oil to the Balkans was also motivated by the war in Yugoslavia. The damage the country suffered was reflected in its oil infrastructure (Interview 01). Because of this, the Russian Federation's PAO Transneft offered to supply oil to the Urinj (Rijeka) refinery. It offered a three-stage project whose capacity would initially be 5 million tonnes, increasing to 10 and 15 million tonnes per year. But only the first variant was realistic in view of the free capacity available in the upstream sections in Hungary and Slovakia. A novel technological issue also surfaced in Croatia: when the JANAF system was being

[5] The following section is based on Vlček (2015, pp. 130–133).

[6] Reverse flow operations are available for the Adria pipeline and in the Gola-Sisak section of the JANAF system.

built, reversing the direction of flow was not part of the plan. The project took advantage of local geographic conditions. From the port of Omišalj, the pipeline travels through the Northern section of the Velika Kapela mountain range in Croatia, which reaches 1400 m above sea level. Coming out of the port, the pipeline walls had to be built thicker to withstand the high pressure created by pumping stations pushing the oil along a route with such high elevation. But the section from the top of the mountains to Sisak did not require such thick walls. Gravity itself did much of the work in this direction. With the direction of flow reversed, however, the thin walls could not withstand the high pressure needed to push the oil up the mountain range—once again the pipeline would burst. Croatia agreed to modernize the pipeline only on condition the Russian Federation provided a bank guarantee. Croatia, that is, needed an obligation for a specific volume of oil based upon which it could take out a loan to modernize JANAF. But Russia offered nothing more than an informal promise; the project never came to fruition.

The stakeholders may not have arrived at an agreement, but they did manage to reveal their diverse interests along the way. Long term, the Russian Federation has tried to diversify its export routes and limit exports via the Druzhba. But its aim in the Balkans is also clearly to control the pipeline and port infrastructure. Croatia is aware of the diversification potential of the oil pipeline as well as the strategic risks ensuing from the Russian Federation's use of the route, risks that arise because reversing the direction of flow largely blocks Central Europe out and damages Croatia's effort to become an EU energy transit country (Socor, 2013). In the end, Croatia definitively turned its back on the negotiations in 2005 and continues to hold a negative view of the project's potential over the long term. The consortium operating the JANAF system, by contrast, generally takes a positive approach to the agreement. Since it is a private company (even if the Government of Croatia is the majority shareholder), it is primarily concerned with maximizing profits, and reversing the direction of flow would bring in new transport revenue.

Russia's interest in acquiring the terminal in Omišalj and gaining control over the route that leads to it is still evident. In June 2013, PAO NK Rosneft signed a Joint Statement of Interest to invest in the energy sector of the Republic of Croatia with the Ministry of Economy of Republic of Croatia. CEO of PAO NK Rosneft Igor Sechin and Minister of Economy of Republic of Croatia Ivan Vrdoljak signed the document at the St. Petersburg International Economic Forum ('*Rosneft Oil Company*',

2013). The document serves as a framework for Russian-Croatian commercial cooperation opportunities. A few months later, in March 2014, Ivan Vrdoljak confirmed meetings with PAO NK Rosneft and PJSC Gazprom Neft over the possible sale of a stake in INA—Industrija nafte d.d. ('*Croatian Side*', 2014; '*Economy Minister*', 2014; 'Газпром нефть', 2014). Both companies are interested in buying the 49.08% share currently held by Hungarian MOL Rt. Should MOL Rt sell these shares, the Croatian government would be willing to sell another 19.8% (Milovan & Žabec, 2013). The current situation is that the Republic of Croatia is interested in a Russian takeover of INA—Industrija nafte d.d. but MOL Rt opposes this. The dispute has been in evidence since 2013, and Hungarian efforts to block the sale are clear. Even though MOL Rt is the single biggest shareholder of INA—Industrija nafte d.d., it does not have management rights. In 2010, Croatian police issued an arrest warrant for MOL Rt's chairman and CEO, Zsolt Hernádi, over accusations he bribed former Croatian Prime Minister Ivo Sanader to gain management rights over INA—Industrija nafte d.d. (in addition to other accusations). Sanader was convicted of corruption in 2012 and the verdict was confirmed by the higher authority in March 2014 ('*Former Croatia*', 2014; Orovic, 2014). He was sentenced to 8.5 years in prison. However, MOL Rt has effectively been blocking the sale of its shares to Russian companies ever since, and PAO NK Rosneft is locked in a dispute with MOL Rt. The president of Croatia has described Ivo Sanader's actions as high treason ('*Cijelu priču*', 2014).

MOL Rt, which owns refineries in Hungary, Slovakia, and Croatia, is among the firms that take Russia's statements on limiting exports over the Druzhba seriously. In September 2011, the company announced it had been working closely with Slovakia's Slovnaft to prepare for the modernization of the Adria Pipeline (i.e. the Gola-Százhalombatta route) ('*MOL, Slovnaft to Invest*', 2011). The project certainly will involve reactivating the pump station, repairing the route, engaging in overall modernization, and increasing pipeline capacity on the Százhalombatta-Šahy route. In May 2012, Slovnaft, Transpetrol, and the MOL Group concluded a Memorandum of Collaboration aimed at modernizing and increasing the capacity of this section ('*Slovnaft, Transpetrol and MOL*', 2012). Although the project is to serve as an alternative, not a replacement to the primary route, the aim is to double capacity. Transpetrol views the project as an opportunity to generate new revenues transporting Russian oil to the Balkan refineries ('*Slovak, Hungarian Firms*',

2012). It must be noted that although the project appears to involve Slovak-Hungarian collaboration, it is actually simply part of MOL's implementation of its business plan. Under growing pressure from statements made by PJSC Gazprom Neft and other Russian companies, MOL is trying to make sure the oil supply to its refineries on the Druzhba Pipeline, its current source, remains secure. The reconstruction was conceived to allow the company to use tankers to transport the entire consumption of Slovnaft (and potentially Százhalombatta) to Omišalj, and from there via JANAF and the Adria all the way to Slovakia (Beer, 2013, p. 42). In February 2015, reconstruction of the Barátság I section between Šahy u Tupé, Slovakia and Tököl, Hungary near Százhalombatta, was completed. The original transportation capacity was not doubled but did increase to 6 million tonnes of oil per year. In addition, two pumping stations located in Hungary were modernized, increasing the transport capacity on the Gola-Százhalombatta section from 6.9 to a maximum 14 million tonnes of oil per year.

Last but not least, among the Russian players in Croatia is OAO Zarubezhneft. The company's plans were in line with PAO NK Rosneft and PAO Transneft plans. In 2012 in Zagreb Nikolai Brunich, former CEO of OAO Zarubezhneft, offered several investment projects including product pipeline from its refinery in Brod, Bosnia and Herzegovina, to Omišalj, Croatia (with the potential to connect it with PJSC Gazprom Neft refineries in Serbia); support over PAO Transneft's ownership entry to JANAF; buying OMV Hrvatska d.o.o.'s 66 petrol stations; and launching exploratory drilling in Slavonia and on the Adriatic shelf ('*Russia's Zarubezhneft*', 2012; Socor, 2012; '*Zarubezhneft Eyes*', 2012). It created Zarubezhneft Adria d.o.o. (a joint venture of OAO Zarubezhneft, 51%; Jadranski Naftovod d.d., 10%; and White Falcon Petroleum Technologies AG from Switzerland, 31%—with the rest being minority shareholders) to participate in tenders for oil exploration and exploitation and planned to create a joint company together with PAO Transneft for the product pipeline project. The Cabinet of Jadranka Kosor was supportive of the proposed projects, but in 2011 a new Cabinet was created after the election with Prime Minister Zoran Milanović. This new government was sceptical about the offers and firmly against letting Jadranski Naftovod d.d. slip out of Croatian hands. OAO Zarubezhneft's situation has thus worsened since 2012. It had no success with any of its investment projects and also withdrew from Zarubezhneft Adria d.o.o. in July 2016.

The Russian strategy in Croatia has been to use Croatia as a starting point for extensive expansion into South-Eastern Europe and Central Europe. Igor Sechin has described Croatia not simply as a market for Russian companies, but instead as a possible base for region-wide operations. He has spoken of a 'breakthrough' to Central Europe, following the country's full entry into the EU (Orovic, 2014; Socor, 2013). To this point, all Russian efforts have gone in vain.

In the meantime, the JANAF-Adria pipeline project has been given European Union's Projects of Common Interest (PCI) status and included in the updated EU PCI list for 2015. The project calls for reconstructing, upgrading, maintaining, and increasing the existing capacity of the JANAF and Adria pipelines. With PCI status, it has become even more difficult for Russian companies to gain access to the ownership structures.

7.3   Research Indicator Assessment

7.3.1   Active Support by Russian State Representatives for Energy Enterprises and Their Activities Abroad

This indicator was confirmed on multiple occasions. Since the beginning of the 2000s, Russian officials have strongly supported the activity of all Russian state-owned oil companies directed at acquiring shares in the JANAF pipeline, reversing the direction of the pipeline, and penetrating the Croatian market. Besides visits by officials of the companies to Zagreb (Nikolai Brunich,[7] Igor Sechin, Alexander Dyukov, Aleksandr Medvedev, and Alexei Miller in 2013), Vladimir Putin himself became involved. He met Croatian President Stjepan Mesić in both Zagreb and Moscow several times, in 2002, 2003, 2007 (within the framework of the Southeast Europe Energy Summit), and 2009. Energy and the economy were always among the topics discussed. Croatian Prime Minister Jadranka Kosor had a meeting with Vladimir Putin in Moscow in 2010 about the oil pipelines in Croatia. Also Russian Energy Minister Sergei Shmatko met his Croatian counterparts (Đuro Popijač, Radimir Čačić) several times, and meetings between other Russian and Croatian officials took place on different occasions.

---

[7] Nikolai Brunich, for example, announced OAO Zarubezhneft's investment projects in Croatia at a news conference at the Russian embassy in Zagreb.

7.3.2   As a Foreign Supplier, Russia Rewards Certain Behaviours and Links Energy Deals to the Client State's Foreign Policy Orientation

No clear evidence was found for this indicator.

7.3.3   Abuse of Infrastructure (e.g. Pipelines) and Differential Pricing to Exert Pressure on the Client State

No clear evidence was found for this indicator.

7.3.4   Efforts to Take Control of the Energy Resources, Transit Routes, and Distribution Networks of the Client State

This indicator is very evident in Croatia. Various Russian state-owned companies, including PJSC Gazprom Neft, PAO NK Rosneft, OAO Zarubezhneft, and PAO Transneft (some working together and supporting each other), have tried to take control of the JANAF pipeline system (and to acquire the Omišalj petroleum port) to reverse its flow from Gola to Omišalj and to export Russian oil to the USA. For example, in the 2000s, PAO NK Rosneft negotiated the direct supply of Russian oil to the USA with Marathon Oil Corporation.

In terms of the retail sector, in 2013 PJSC Gazprom Neft tried unsuccessfully to purchase the smaller petrol station networks of OMV Hrvatska d.o.o. and Crodux derivati dva d.o.o. PAO Lukoil was partially successful with its strategy of becoming a competitor of Croatian INA—Industrija nafte d.d. On the one hand, it managed to gain entry into the Croatian market by purchasing the petrol station network of Europa Mil d.o.o. and Crobenz d.d. and building its own petrol stations; on the other, development has been slow and currently the company has met only a third of its 2011 projections.

7.3.5   Disruption (by Various Means) of Alternative Supply Routes/Sources of Supply

Russian companies are not shareholders in any section of the oil transport routes leading to Croatia, and Russian oil plays a minority role in the oil import portfolio. Russian export blend is a minority oil type refined only in Sisak, with just 0.58 mta of imports in 2015. No clear evidence was therefore found for this indicator.

7.3.6   Efforts to Gain a Dominant Market Position in the Client Country

The efforts of Russian state-owned companies are directed at taking control of the JANAF pipeline system and the Omišalj petroleum port for purposes of exporting Russian oil overseas, as well as becoming a dominant market player in Croatia and South-Eastern Europe. The Russian strategy in Croatia has been to use Croatia as a starting point for extensive expansion into South-Eastern Europe and Central Europe. Igor Sechin has described Croatia not simply as a market for Russian companies, but instead as a possible base for region-wide operations. He has spoken of a 'breakthrough' to Central Europe, following the country's full entry into the EU (Orovic, 2014; Socor, 2013). To this point, however, all of Russia's efforts have gone in vain.

7.3.7   Efforts to Eliminate Competitive Suppliers

There is no evidence to clearly confirm this indicator. Potentially taking control of the JANAF pipeline system and reversing the flow of oil could block out Central Europe. This could eliminate supplies to Central Europe; however, this has never come to pass. Potentially eliminating competitive suppliers by controlling the JANAF pipeline could theoretically be seen in Bosnia and Herzegovina (OAO Zarubezhneft's Brod refinery) and Serbia (at PJSC Gazprom Neft's refineries in Novi Sad and Pančevo). The Russian Federation would own both the refineries and the pipeline route, which would allow it to block out alternative suppliers. This is, however, purely theoretical: Russian companies have not succeeded in acquiring stock in the JANAF pipeline's shareholders.

7.3.8   Acting Against Liberalization

No clear evidence was found for this indicator.

7.3.9   Diminishing the Importance and Influence of Multilateral Regimes Such as the EU

No clear evidence was found for this indicator; however, the frequency of meetings involving Russian officials (or the management of Russian state-owned companies) dropped considerably after Croatia's entry into the

European Union in July 2013. It is clear that activity by Russian representatives aimed at penetrating the Croatian energy sector rose sharply once
the exact date of the country's accession was made public: Nikolai Brunich
visited Zagreb in January 2013; Alexander Dyukov, Aleksandr Medvedev,
and Alexei Miller in January and February 2013; Igor Sechin in June 2013.

7.3.10   Attempts to Control the Entire Supply Chain (Regardless of Commercial Rationale)

PAO NK Rosneft and PJSC Gazprom Neft's interest in purchasing the
MOL Rt's share in INA—Industrija nafte d.d. was part of their strategy.
The port of Omišalj and INA—Industrija nafte d.d.'s refining capacities
were part of the strategy for exporting oil and oil products overseas. All
the evidence, however, suggests that these efforts should be understood as
commercial. No attempt was made to simply take over the supply chain
and thereby control the market within Croatia's borders; the efforts have
always targeted the use of Croatian infrastructure for overseas exports.

7.3.11   Economically Irrational Steps Taken to Maintain a Particular Position in the Client State's Market

No clear evidence was found for this indicator (Table 7.3).

Table 7.3 Summary of indicators

Indicator	Found	Found with
Active support by Russian state representatives for energy enterprises and their activities abroad	Yes	PAO Lukoil, PJSC Gazprom Neft, PAO NK Rosneft, OAO Zarubezhneft, PAO Transneft
As a foreign supplier, Russia rewards certain behaviours and links energy deals to the client state's foreign policy orientation	No	–
Abuse of infrastructure (e.g. pipelines) and differential pricing to exert pressure on the client state	No	–
Efforts to take control of the energy resources, transit routes, and distribution networks of the client state	Yes	PJSC Gazprom Neft, PAO NK Rosneft, OAO Zarubezhneft, PAO Transneft

(continued)

Table 7.3   (continued)

Indicator	Found	Found with
Disruption (by various means) of alternative supply routes/sources of supply	No	–
Efforts to gain a dominant market position in the client country	Yes	PJSC Gazprom Neft, PAO NK Rosneft, OAO Zarubezhneft
Efforts to eliminate competitive suppliers	Inconclusive	–
Acting against liberalization	No	–
Diminishing the importance and influence of multilateral regimes such as the EU	Inconclusive	–
Attempts to control the entire supply chain (regardless of commercial rationale)	Inconclusive	PJSC Gazprom Neft, PAO NK Rosneft
Economically irrational steps taken to maintain a particular position in the client state's market	No	–

Source: Author

Note: *Inconclusive* means some indications of this behaviour were found, but not of a shape, size, or importance to be ascribed to strategic behaviour and thus fulfil the indicator

SOURCES

Agencija za ugljikovodike. (n.d.). Retrieved from http://www.azu.hr/

Beer, G. (2013). Transpetrol s nožom na krku. Štátnému prepravcovi ropy prestáva tiect biznis do náručia. *Trend, 2013*(1), 42–43.

Cijelu priču mogu okarakterizirati samo jednom rječju, a to je veleizdaja. (2014, June 14). *Vijesti.hr*. Retrieved from http://www.vijesti.rtl.hr/novosti/1218195/cijelu-pricu-mogu-okarakterizirati-samo-jednom-rjecju-a-to-je-veleizdaja/

Contracts Signed with INA, Vermilion Energy on Onshore Hydrocarbon Exploration and Exploitation. (2016, June 10). *Europe Balkan Latest News*. Retrieved from https://eblnews.com/news/business/contracts-signed-ina-vermilion-energy-onshore-hydrocarbon-exploration-and-exploitation-24644

Croatia Signs Contracts for Onshore Oil and Gas Exploration. (2016, June 10). *Reuters*. Retrieved from http://www.reuters.com/article/croatia-crude-idUSL8N1921GT

Croatian Energy Regulatory Agency. (2014). *Annual Report 2013*. Retrieved from https://www.hera.hr/en/docs/HERA_Annual_Report_2013.pdf

Croatian Side in Rosneft, Gazprom Neft Meetings, Mulls No INA Exit – Media. (2014, March 14). *SeeNews*. Retrieved from https://seenews.com/news/croatian-side-in-rosneft-gazprom-neft-meetings-mulls-no-ina-exit-media-409710

Crodux Could be Sold to Gazprom Neft. (2016, May 10). *Balkan Energy*. Retrieved from http://balkanenergy.com/crodux-could-be-sold-to-gazprom-neft-croatia-10-may/

Economy Minister Confirms Talks with Rosneft, Gazprom Neft on INA. (2014, March 14). *dalje.com*. Retrieved from http://arhiva.dalje.com/en-croatia/economy-minister-confirms-talks-with-rosneft-gazprom-neft-on-ina/502745

Former Croatia PM Ivo Sanader Convicted of Corruption. (2014, March 11). *BBC News*. Retrieved from http://www.bbc.com/news/world-europe-26533990

'Газпром нефть' намерена участвовать в тендерах на шельфе Адриатики ['Gazprom Neft' Plans to Participate in Tenders for the Adriatic Shelf]. (2014, April 7). *Ведомости [Vedomosti]*. Retrieved from http://www.vedomosti.ru/business/news/2014/04/07/gazprom-neft-namerena-uchastvovat-v-tenderah-na-shelfe

Gazprom Neft Without an Oil and Gas Exploration Licence in Croatia. (2015, June 8). *WiseBroker*. Retrieved from http://wisebroker.rs/niis-gazprom-neft-without-an-oil-gas-exploration-licence-in-croatia/

Gazprom Neft to Buy Former OMV Gas Station Network from Crodux, Regional Expansion on the Way. (2013, July 11). *Serbia Energy*. Retrieved from http://serbia-energy.eu/croatia-gazpromneft-to-buy-former-omv-gas-station-network-from-crodux-regional-expansion-on-the-way/

INA—Industrija nafte d.d. (2009). *INA Annual Report 2009*. Retrieved from http://www.ina.hr/UserDocsImages/g_izvjesca_pdf/gi_2009_eng.pdf

INA—Industrija nafte d.d. (2014). *Annual Report 2014. Economic, Social and Environmental Performance*. Retrieved from http://www.ina.hr/UserDocsImages/INA_eng_Annual_Report_final.pdf

INA—Industrija nafte d.d. (2015). *Annual Report 2015. Economic, Social and Environmental Performance*. Retrieved from http://www.ina.hr/UserDocsImages/investitori/Godi%C5%A1nje%20izvje%C5%A1%C4%87e%202015_ENG_final_objava.pdf

International Energy Agency. (n.d.). Retrieved from https://www.iea.org/

International Energy Agency. (2011). *Energy Policies of IEA Countries – Hungary – 2011 Review*. Paris: IEA Publications. Retrieved from http://www.iea.org/publications/freepublications/publication/hungary2011_web.pdf

Jadranski Naftovod d.d. (n.d.). Retrieved from http://www.janaf.hr/

'Lukoil Wants 100 Gas Stations in Croatia'. (2008, June 23). *Nacional*. Retrieved from http://arhiva.nacional.hr/en/clanak/46826/lukoil-wants-100-gas-stations-in-croatia

Lukoil to Buy 150 Petrol Stations in Croatia. (2008, November 18). *dalje.com*. Retrieved from http://arhiva.dalje.com/en-economy/lukoil-to-buy-150-petrol-stations-in-croatia/204838

Marathon Oil and OMV Back Down on Drilling Plans Offshore Croatia. (2015, July 29). *Offshore Energy Today*. Retrieved from http://www.offshoreenergytoday.com/marathon-oil-and-omv-back-down-on-drilling-plans-offshore-croatia/

Milovan, A., & Žabec, K. (2013, November 26). Jutarni doznaje: MOL pregovara s Rusima. *Jutarnji Vijesti*. Retrieved from http://www.jutarnji.hr/vijesti/hrvatska/jutarnji-doznaje-mol-pregovara-s-rusima-madari-sve-blize-prodaji-udjela-ine/923168/

MOL Group. (n.d.). Retrieved from https://molgroup.info/

MOL, Slovnaft to Invest EUR80 Million in Adria Oil Pipeline-Agency. (2011, September 13). *Dow Jones News*. Retrieved from http://www.advfn.com/news_MOL-Slovnaft-To-Invest-EUR80-Million-In-Adria-Oil_49158149.html

Orban, A. (2008). *Power, Energy, and the New Russian Imperialism*. Westport and London: Praeger Security International.

Orovic, J. (2014, February 26). Russia's Rosneft Said Looking to Buy INA and Petrol in the Balkans. *bne IntelliNews*. Retrieved from http://www.intellinews.com/russia-s-rosneft-said-looking-to-buy-ina-and-petrol-in-the-balkans-500000192/?archive=bne

Pavlic, V. (2016, July 28). Croatian Members of INA Board of Directors Prevent Closure of Refinery in Sisak. *Total Croatia News*. Retrieved from http://www.total-croatia-news.com/item/13302-croatian-members-of-ina-board-of-directors-prevent-closure-of-refinery-in-sisak

Rosneft Oil Company and Ministry of Economy of Republic of Croatia Signed 'Joint Statement of Interest to Invest in the Energy Sector of the Republic of Croatia'. (2013, June 21). *Rosneft Press Release*. Retrieved from https://www.rosneft.com/press/releases/item/114370/

Russia's Zarubezhneft Eyes Gas Stations in Croatia, Bosnia. (2012, June 12). *Expatica*. Retrieved from http://www.expatica.com/ru/news/Russias-Zarubezhneft-eyes-gas-stations-in-Croatia-Bosnia_344691.html

Russians to Enter Ina Instead of MOL. (2007, June 19). *Nacional*. Retrieved from http://arhiva.nacional.hr/en/clanak/35630/russians-to-enter-ina-instead-of-mol

Slovak, Hungarian Firms Sign Accord on Increasing Capacity of Oil Pipeline. (2012, December 8). *Hospodářské noviny*.

Slovnaft, Transpetrol and MOL Have Signed Memorandum on Cooperation in Connection with Modernisation and Increasing of Transport Capacity for Pipeline Friendship 1 Between Slovakia and Hungary. (2012, May 12). *Press Report of Slovnaft, a.s.* Retrieved from http://killajoules.wikidot.com/blog:3510

Socor, V. (2010). Russia and Croatia Resurrect Druzhba-Adria Oil Transport Scheme. *Eurasia Daily Monitor*, 7(45). Retrieved from https://jamestown.org/program/russia-and-croatia-resurrect-druzhba-adria-oil-transport-scheme/

Socor, V. (2012, January 30). Russian Oil Business Targeting EU's Entrant Croatia. *Eurasia Daily Monitor*, 9(20). Retrieved from https://jamestown.org/program/russian-oil-business-targeting-eus-entrant-croatia/

Socor, V. (2013). Igor Sechin Door-Crashing in Croatia. *Eurasia Daily Monitor*, 10(116). Retrieved from https://jamestown.org/program/igor-sechin-door-crashing-in-croatia/

Veselica, L. (2015, August 2). Adriatic Oil, Gas Exploration Raises Concerns for Croatia Tourism. *Phys.org*. Retrieved from http://phys.org/news/2015-08-adriatic-oil-gas-exploration-croatia.html

Vlasnik Croduxa, umirovljeni general Ivan Čermak, ne prodaje svoj biznis u Hrvatskoj Rusima. (2016, May 11). *Poslovni dnevnik*. Retrieved from http://www.poslovni.hr/domace-kompanije/vlasnik-croduxa-umirovljeni-general-ivan-cermak-ne-prodaje-svoj-biznis-u-hrvatskoj-rusima-312739

Vlček, T. (2015). *Alternative Oil Supply Infrastructures for the Czech Republic and Slovak Republic* (1st ed.). Brno: Masarykova univerzita.

Zarubezhneft Eyes Retail Assets in Croatia. (2012, January 25). *Sputnik News*. Retrieved from https://sputniknews.com/business/20120125170946286/

LIST OF INTERVIEWS

Interview 01: Bratislava, Slovakia, August 25, 2014.

CHAPTER 8

Greece

8.1   Crude Oil Sector General Information

8.1.1   Introduction and Upstream

Greece (officially the Hellenic Republic) is an EU member country for which oil is a dominant energy source, accounting for around 45% of its total primary energy supply (TPES) in 2012, rising to 55% in 2014 (International Energy Agency, 2015, p. 218; Mearns, 2015). Greece's domestic oil reserves are minimal, reflecting the negligible domestic production of just 136 tonnes in 2014. These reserves will allow Greece to extract oil for another 22 years at the current reserve/production ratio (Eni SpA, 2015, p. 16). The country is thus nearly completely dependent on external sources of oil. The members of the Organization of the Petroleum Exporting Countries (OPEC) such as Iran, Libya and Saudi Arabia, and the former Soviet Union (FSU) have served as major import sources for crude oil. Of the FSU countries, the Russian Federation forms the largest source for crude oil, accounting for between 33% and 40% of the total supply (International Energy Agency, 2011, p. 55; Wattles,

---

This chapter partially builds on and develops a paper entitled 'Russia's energy relations in Southeastern Europe: an analysis of motives in Bulgaria and Greece' (Jirušek, Vlček, & Henderson, 2017), published in the Taylor & Francis journal *Post-Soviet Affairs* (www.tandfonline.com).

© The Author(s) 2019
T. Vlček, M. Jirušek, *Russian Oil Enterprises in Europe*,
https://doi.org/10.1007/978-3-030-19839-8_8

2015). In 2014, roughly 4.7 million tonnes of crude oil were imported from the Russian Federation.

Greece's crude oil imports from the Russian Federation, which have reached as high as 40% in recent years, formerly hovered around the 20% level. It was the United Nations economic sanctions against Iran from 2010 that brought the change. More than 50% of the oil Greece imported in 2011 was of Iranian origin (Onti, 2013), and the sanctions therefore sharply impacted the structure of Greek imports. Imports from post-Soviet countries have risen since 2011 (especially from Kazakhstan); they have nearly doubled if one speaks of the Russian Federation itself. This situation seems to be returning to pre-sanction values since the sanctions against Iran were lifted in 2015. A short time after this, Hellenic Petroleum S.A. announced that Greece would acquire about a quarter of its needs from Iran under a long-term supply agreement with the National Iranian Oil Company (NIOC) (Eqbali, 2016). One motivation for reacting so quickly to the lifting of sanctions was Hellenic Petroleum S.A.'s debt of US $755 million (US $800 million including the debts of other Greek companies), which was settled during the negotiation of the new long-term supply agreement ("*Iran, Greece Seal*", 2016; Interview 04; Interview 06); another was Greek shipowners' and shipbuilders' interest in doing business with Iran (Table 8.1).

Domestic production takes place in the Prinos-Kavala basin located offshore in the Gulf of Kavala. Extraction is being carried out at three oil fields—Prinos, Prinos North, and South Kavala—with successful exploratory activity being conducted at the Epsilon oil field in the same basin, the Ioannina oil field in Eastern Greece, and the Katakolon oil fields on the Eastern Peloponnese Peninsula. The only crude oil-producing company in the country is Energean Oil & Gas S.A., owned

Table 8.1 Greece crude oil data

	2010	2012	2013	2014
Consumption	18,524	15,835	14,142	14,092
Production	273	273	136	136
Import dependency	100%	100%	100%	100%

Source: Eni SpA, 2015, pp. 9–21

Note: Figures in thousands of tonnes annually, conversion by author

by private founding shareholders[1] (55%), and Third Point LLC, an investment adviser based in New York (45%) (Energean Oil & Gas S.A., n.d.).

8.1.2   Midstream

Greece is essentially fully dependent on external sources of oil and imports crude oil exclusively over maritime routes. Apart from the pipeline to Former Yugoslav Republic of Macedonia (FYROM), there are no international crude oil pipelines in Greece. There are ten maritime oil terminals in the country, six of which can accept crude oil (four located near refineries); they are Aspropyrgos, Eleusis, Thessaloniki, Aghioi Theodori, Pachi (Megara), and Agia Trias (International Energy Agency, 2010, p. 7). The sole pipeline extends 220 km from the Thessaloniki oil terminal to OKTA Crude Oil Refinery A.D. in Skopje, FYROM. The OKTA refinery was closed in 2013, and it was agreed it would be used as a main hub for the distribution of oil products. Hellenic Petroleum S.A. could now import products from Serbian Nafta industrija Srbije a.d. at a lower cost than if they were produced in OKTA A.D. (Interview 01). This also impacted the crude oil pipeline that, as noted, was converted into an oil product pipeline (Littler, Pitsas, & Jørgensen, 2015, pp. 71–72). OKTA A.D. is owned by EL.P.ET. Balkaniki S.A.[2] (81.51%), Pucko Petrol (10.87%), and the Company employees (7.62%) (Hellenic Petroleum S.A., n.d.).

8.1.3   Downstream

There are four refineries in Greece located in Elefsina (Athens), Aspropyrgos, Korinthos, and Thessaloniki. Together they have a total refining capacity of 26.3 million tonnes annually (mta; see Table 8.2). This total capacity is roughly the double of Greece's annual demand, meaning Greece is a net exporter of petroleum products. Exports go mainly to nearby non-EU countries, in particular Turkey, Lebanon, Libya, FYROM, and Gibraltar, as well as to more distant Singapore (Danchev & Maniatis, 2014, p. 31).

---

[1] One of the shareholders is Prime Marine, a Greek maritime transportation company specializing in crude oil, chemical products, and natural gas shipping, owned by one of the founding shareholders.

[2] EL.P.ET. Balkaniki S.A. is a company jointly controlled by Hellenic Petroleum S.A. (63%) and Consortium of banks APE (37%) (OKTA Crude Oil Refinery A.D., 2014, p. 13).

Table 8.2   Capacity of Greek refineries

Refinery	Owner	Refining capacity (mta)	Type	Year established
Aspropyrgos	Hellenic Petroleum S.A.	7.5	Cracking (FCC)	1958
Elefsina (Eleusis)	Hellenic Petroleum S.A.	5.0	Hydrocracking	1972
Thessaloniki	Hellenic Petroleum S.A.	4.5	Hydroskimming	1966
Korinthos (Aghioi Theodori)	Motor Oil Hellas S.A.	9.3	Cracking (FCC)	1972

Source: Danchev & Maniatis, 2014, p. 26; International Energy Agency, 2011, p. 58; adjusted by author

Hellenic Petroleum S.A. controls 65% of the refining sector in Greece and 30% of the retail. Refining is the company's core business. It owns a domestic network of 1716 petrol stations (through its EKO and HELLENIC FUELS subsidiaries), and a network of 287 petrol stations abroad, making it one of the key fuel-marketing players in Cyprus, Serbia, Bulgaria, and Montenegro as well. Hellenic Petroleum S.A. is also active in the power and gas sectors. The Greek state controlled 35.48% of company assets as of 2014. Other shareholders are Paneuropean Oil & Industrial Holdings S.A. (42.57%), Greek institutionals (8.5%), international institutionals (5.35%), and private investors (8.09%) (Hellenic Petroleum S.A., n.d.). Paneuropean Oil & Industrial Holdings S.A., based in Luxembourg, operates as a subsidiary of the Latsis Group,[3] an Athens-based company active in banking, civil engineering, energy, water and air transport, real estate, oil, and shipping through its many subsidiaries. Hellenic Petroleum S.A. also owns and operates the crude oil pipeline to FYROM.

Motor Oil Hellas S.A. controls the remaining 35% of the refining sector in Greece through its Korinthos refinery. It operates a network of approximately 450 retail petrol stations in Greece under the AVIN brand, which gives it an approximate 10% share in the domestic market. Its shareholder structure includes US-based Petroventure Holdings Ltd.[4] (40%), Doson Investments Company (7.8%), and free float shares (52.2%) (Motor Oil Hellas S.A., n.d.).

---

[3] The Latsis family is among the most prominent Greek business families.

[4] This company belongs to the prominent Greek business family Vardinoyannis (Michaletos, 2011).

There are around 20 companies that are active on the Greek oil products retail market. In 2009–10, the two biggest (besides Hellenic Petroleum S.A. and Motor Oil Hellas S.A.), British Petrol and Shell, sold their retail networks to Hellenic Petroleum S.A. (BP), and Motor Oil Hellas S.A. (Shell). These transactions helped the respective Greek companies to achieve strong positions in the domestic market. Russian companies are not present in Greece's petroleum retail sector, with the exception of the Greek office of Lukoil Marine Lubricants Ltd., a company active in the worldwide maritime lubricant market.

8.2 Recent Market Developments and Russian Activities

There is a long history of positive relations between the Russian Federation and Greece that continue to this today. Some even portray Greece (and Cyprus) as a Russian Trojan Horse within the EU, as the connections to Russia have both ancient cultural and more recent geopolitical and economic roots (Leonard & Popescu, 2007, p. 27). Russia-Greece relations are based primarily on military cooperation: Russia supports dealings with Turkey and other Greek political interests, and of course the energy sphere; and Greece promotes various Russian interests within the EU (Hegedüs, 2010, p. 3). Greece is also important to Russia because of its relationship to the Balkan countries and its regional geopolitical position.

But these relations are often overstated and misunderstood. While it is true that Greece is interested in a good relationship with the Russian Federation given Russia's investments in sports, real estate, tourism, and the economic residence programme, it is also the case that the two countries' close relations are prompted by the need for a Russian counterbalance to Turkey, Russia's backing on the Cyprus issue for Greece, a common religion, and cultural and historical proximity (the Russian Empire's assistance in the Greek War of Independence from Ottoman control in the nineteenth century might be the best example) (Interview 01; Interview 06). One must distinguish, though, between reality and the image that Greece wishes to create. For Greece is in fact a pro-Western country, as a look at its politics in practice makes clear,[5] and its relations with Russia are

---

[5] For example, the EU sanctions against Russia were never vetoed by Greece and all the official EU statements during the Ukraine crisis were officially supported by the Greek government.

based on diplomatic gestures of goodwill and the behaviour and state-ments of individual politicians intent on securing good relations with both the EU and Russia. It has never been the Russian Federation's plan to spend any political capital in Greece; rather it has sought support for issues elsewhere (i.e. sanctions against Russia, development in Ukraine, etc.) (Interview 01; Interview 02; Interview 04). Of the Balkan countries, Russia places its focus on Bulgaria, Serbia, and Montenegro, rather than on Greece.

Russia is viewed very positively by the Greek public, something that might be related to Greece's economic problems and Vladimir Putin's macho optics. The policy of promoting good relations with Russia is also a somewhat easy sell for Greek politicians and is thus used to seek support for domestic political parties. The public's acceptance and support of rela-tions with Russia is also a consequence of Greece's complicated relations with the EU in the light of the Greek government-debt crisis (Interview 02).

Greece is a long-term supporter of a planned crude oil pipeline running from Burgas in Bulgaria to Alexandroupolis in Greece. The primary logic of this pipeline is to bypass the Bosporus Strait and serve as an alternative route for Russian and Caspian oil not only to Greece, Bulgaria, and Macedonia but to the global market through the Alexandroupolis oil ter-minal. The project is the shortest of many routes[6] that would bypass the Bosporus Strait, where navigation accidents are frequent due to high traf-fic and difficult navigation. The Russians would also like to see it work to avoid having to transport oil through Turkish territory, which is a liability similar to what occurs with the export of Russian hydrocarbons through Ukraine.

The plans to construct the Burgas-Alexandroupolis pipeline date back to 1994 when the first Memorandum of Cooperation between the Russian Federation, Greece, and Bulgaria was signed. The 2007 plan spoke of a 279 km oil pipeline with a total budget of EUR 0.9–1 billion (later revised to EUR 1.5 billion) and an initial capacity of 35 mta, with potential extension to 50 mta (Jirušek, Vlček, & Henderson, 2017, p. 346; Ministry of Development of the Hellenic Republic, 2007, p. 4). Numerous high-level summits aimed at intensifying cooperation have been held since 1994, and a number of cooperation memoranda were signed. In 2007,

---

[6] Such as the Odessa-Brody-Płock pipeline from Ukraine to Poland, the Constanţa-Trieste from Romania to Italy, the Burgas-Vlorë from Bulgaria to Albania (also known as AMBO), or two Turkish options: Kiyikoy-Ibrice and Samsun-Ceyhan.

Table 8.3 Shareholder structure of Trans-Balkan Pipeline B.V.

Country	Share (%)	Company	Company shareholders
Russian Federation	51.0	Pipeline Consortium Burgas-Alexandroupolis Ltd.	OAO AK Transneft (33.34%), OAO NK Rosneft (33.33%), PAO Gazprom Neft (33.33%)
Republic of Bulgaria	24.5	JSC Project Company Oil Pipeline Burgas-Alexandroupolis—BG AD	Ministry of Finance of the Republic of Bulgaria
Hellenic Republic	23.5	Helpe-Thraki A.E.	Hellenic Petroleum S.A. (25%), Thraki S.A. (75%)
Hellenic Republic	1.0	Hellenic Republic	Government of the Hellenic Republic

Source: Prometheus Gas S.A., n.d.; Trans-Balkan Pipeline B.V., n.d.; adjusted by author

Note: Thraki S.A. is a subsidiary of Prometheus Gas S.A., a Greek-Russian joint venture in which OAO Gazprom through its 100% subsidiary OOO Gazpromexport controls 50% of the equity of Prometheus Gas S.A. The other 50% is controlled by Dimitrios Ch. Copelouzos

the international project company Trans-Balkan Pipeline B.V. was created; its shareholder structure is depicted in Table 8.3.

The reason the pipeline has still not been built is Bulgaria: the government officially threatened to abandon the project because of its environmental risks and eventually did just that. However, an expert source has it that the environmental reasoning was not accurate, and a key reason for abandoning the project was that Bulgaria's Western allies did not support it.

In reality, the project lost its political backing in Bulgaria after 2009 when a new centre-right government took office. Since that time, Bulgarian representatives have written the project off regularly, even though Greece and the Russian Federation were still in favour. Since 2011, the project has been put on hold by Russia.

There are several explanations for Bulgarian politics (Papaspanos, 2010); one is that the country's about-turn was geopolitical, that is, it reflected a desire not to block any potential new route from the Caspian region that would carry non-Russian oil to the EU. Construction of the Burgas-Alexandroupolis pipeline would allow easier export of Russian oil over the Caspian and thus negatively affect plans to develop transport routes for Caspian oil. By controlling a new export route on the Balkan Peninsula, the Russians would be able to better control the competing Caspian resources and secure their geopolitical position. In fact, there was an agreement between Russia and Kazakhstan to use Russian oil in the Caspian Pipeline Consortium (CPC) for Russian refineries in the Balkans— this at a time when the Burgas-Alexandroupolis pipeline was still expected to materialize (Interview 02).

The complicated course of development of Balkan oil and gas projects also stems from the fact that the Balkan countries understand their geopolitical strength and attractiveness for EU energy security measures. According to Venelin Tsachevsky, since the gas crisis of 2009, the Balkan region has been considered an important alternative route for bypassing energy supplies from Russia (Tsachevsky, 2011, pp. 5–6). Those Balkan countries that are chosen as transit countries for new oil and gas pipeline projects from the Caspian region would benefit considerably from their position. This is one of the main reasons for such strong competition over oil and gas projects in the Balkan Peninsula.

Turkey, which has its own alternative projects to bypass the Bosporus and Dardanelles Straits and to get Caspian oil to the EU market, is in a difficult position with regard to the Burgas-Alexandroupolis pipeline. According to Konstantin Kalinkov, the pipeline would, on the one hand, compete with Turkey's interests in being the sole monopolist regional centre for the transit of energy resources; on the other, it would solve the crucial problem of constantly increasing tanker traffic through the Bosporus (Kalinkov, 2010, p. 66). Even though Turkey's pronouncements on the issue are overstated and intended to serve Turkish interests, the navigation issues in the Straits are real (Interview 06). Tankers and other ships sail very close around the Turkish capital and traffic is extensive; it is quite normal for ships to wait for days to pass through the Straits.

In 2012, Bulgaria's finance minister, Simeon Djankov, had described the project as 'economically nonviable' (Sideris, 2015) and in 2013 Bulgaria renounced the 2007 agreement between the three countries involved in the Burgas-Alexandroupolis pipeline. None of the other participants have officially quit the project so far. Since 2015, though, US-based Chevron Corporation, which is a member of the Caspian Pipeline Consortium, has been holding talks with Bulgaria, as Chevron has shown interest in reviving the project to allow for the export of Caspian oil after expansion of the CPC. The frozen Burgas-Alexandroupolis pipeline project might thus still have some future, but it can hardly be realized without foreign investment. The pipeline is uncompetitive with the current low oil prices, as its planned operational costs were high.

8.3   Research Indicator's Assessment

8.3.1   Active Support by Russian State Representatives for Energy Enterprises and Their Activities Abroad

This indicator has been detectable several times in the course of Greek-Russian relations, especially as regards Russian support of the Burgas-Alexandroupolis pipeline project in the 1990s and 2000s. Vladimir Putin, together with President Sergei Stanishev of Bulgaria and Prime Minister Kostas Karamanlis of Greece, took part in a summit meeting in Athens in 2007, where the three leaders signed an intergovernmental agreement (Tziampiris, 2010, p. 80). Former Industry and Energy Minister Viktor Khristenko signed the document on Russia's behalf. The agreement establishing the international project company Trans-Balkan Pipeline B.V. was signed in Moscow at the end of 2007. Vladimir Putin gave verbal support to the project on various occasions, as did Viktor Khristenko, before the signing took place (Jirušek et al., 2017, p. 346).

In 2002–03, Russia's PAO Lukoil (together with the Latsis Group) expressed a desire to buy Hellenic Petroleum S.A., and the Russian government greatly supported the effort. Vagit Y. Alekperov, CEO of PAO Lukoil, was strongly in favour of the project, and Vladimir Putin's visit to Greece in 2001 was associated with this objective. PAO Lukoil's interests in Southern Europe expanded at that time after asset purchases and purchase efforts were made in other Balkan countries (Romania, Bulgaria).

However, as Inna Gaiduk and Oleg Lukin have indicated, Athens decided not to sell Hellenic Petroleum S.A. to PAO Lukoil in early 2003, stating that the proposal by PAO Lukoil and Latsis Group was 'unacceptable from the point of view of national interests' (Gaiduk & Lukin, 2006, cited in Poussenkova, 2012, p. 194). Since that time, no effort to take control of Greek resources or infrastructure has been made.

8.3.2   As a Foreign Supplier, Russia Rewards Certain Behaviours and Links Energy Deals to the Client State's Foreign Policy Orientation

Due to the specific nature of the global oil market, it is difficult or impossible to connect oil prices to a customer's foreign policy orientation. The Russian share of Greek crude oil imports is around 20%, and long-term oil contracts are usually concluded for just one- or two-year periods. But signs of special relations between Athens and Moscow have been visible outside the energy sector. Quite recently, the EU discussed potentially prolonging the sanctions against Russia. In January 2015, Prime Minister Alexis Tsipras stated that Greece would not consent to the prolongation.

Greece has been perceived to be a Russian Trojan Horse inside Europe in view of its track record of supporting Russian policy (especially, but not only, the ruling party Syriza). Syriza supported Russia in the Ukraine crisis, in the Crimea crisis (Alexis Tsipras supported the Crimean referendum), and during the first round of Western sanctions (*"New Greek Government"*, 2015). Greek officials have made a number of visits to Moscow in recent years. Pavel Baev understands this to be a morale boost for Russia. With the considerable economic problems that followed the sanctions, Russian propaganda painted the EU as morally corrupt and politically divided, and Greece's failure to comply with the EU policy supports this view (Baev, 2015).

8.3.3   Abuse of Infrastructure (e.g. Pipelines) and Differential Pricing to Exert Pressure on the Client State

There is no imported crude oil pipeline in Greece, and the Russian share of Greek crude oil imports is usually around 20%. The existence of several oil terminals in Greece and a diversified oil import portfolio makes it structurally difficult to exert influence over infrastructural bottlenecks.

8.3.4   Efforts to Take Control of the Energy Resources, Transit Routes, and Distribution Networks of the Client State

The Burgas-Alexandroupolis pipeline was meant to be the almost exclusive conduit for Russian oil; the Russians thus sought the vast majority of equity shares in the pipeline, while the Greeks wanted to limit the Russians to 51% and divide the remaining equity with Bulgaria (*"Putin and Purvanov"*, 2006). This is what eventually happened, and the Russian company Pipeline Consortium Burgas-Alexandroupolis Ltd. now has 51% of the equity shares in Trans-Balkan Pipeline B.V.

In 2002–03, Russia's PAO Lukoil (together with the Latsis Group) expressed a desire to buy Hellenic Petroleum S.A., and the Russian government greatly supported the effort. Vagit Y. Alekperov, CEO of PAO Lukoil, was strongly in favour of the project, and Vladimir Putin's visit to Greece in 2001 was associated with this objective. PAO Lukoil's interests in Southern Europe expanded at that time after asset purchases and purchase efforts were made in other Balkan countries (Romania, Bulgaria).

However, as Inna Gaiduk and Oleg Lukin have indicated, Athens decided not to sell Hellenic Petroleum S.A. to PAO Lukoil in early 2003, stating that the proposal by PAO Lukoil and Latsis Group was 'unacceptable from the point of view of national interests' (Gaiduk & Lukin, 2006, cited in Poussenkova, 2012, p. 194). After this failure PAO Lukoil unsuccessfully tried to purchase Motor Oil Hellas S.A. in 2005–06 (Interview 01). Since that time, no effort to take control of Greek resources or infrastructure has been made.

8.3.5   Disruption (by Various Means) of Alternative Supply Routes/Sources of Supply

There is no imported crude oil pipeline in Greece, and the Russian share of Greek crude oil imports is around 20%. The existence of several oil terminals in Greece and a diversified oil import portfolio makes it structurally difficult to exert influence over infrastructural bottlenecks.

8.3.6   Efforts to Gain a Dominant Market Position in the Client Country

Such efforts were visible in the late 1990s–early 2000s in relation to the failed Hellenic Petroleum S.A. acquisition. While PAO Lukoil planned to enter the Balkan market, in recent years the exact opposite has come to pass, with the traditional Russian oil companies in the process of withdrawing from Central and Eastern Europe (CEE) and SE markets. In 2014, PAO Lukoil's subsidiaries sold their petrol stations in the Czech Republic (44 stations), Hungary (75 stations), Slovakia (19 stations), and Ukraine (240 stations). In 2015, 37 stations were sold in Estonia, and in 2016 the process continued with networks sold in Poland, Lithuania, and Latvia (all together approximately 230 stations) ("*Lukoil wycofuje*", 2016; "*Olerex to Purchase*", 2015; "*Time for Mergers*", 2015). PAO Lukoil CEO Alekperov explained in a December 2015 interview on Rossiya-24 TV in Moscow that the move was made due to rising anti-Russian sentiment in these countries: 'We're facing challenges in a number of countries, including Ukraine, where we were forced to sell our assets though they were among the best' ("*Russia's Lukoil*", 2015). It represented the company's forced reaction to market developments and consumer boycotts of PAO Lukoil petrol stations after the annexation of Crimea by the Russian Federation on 18 March 2014.

8.3.7   Efforts to Eliminate Competitive Suppliers

This indicator is not relevant to Greece, as no Russian companies are present in the Greek market. Also, Russia is by far not the dominant supplier of crude oil to Greece; Greece has a diversified oil import portfolio.

8.3.8   Acting Against Liberalization

This indicator is not relevant to Greece, as no Russian companies are present on the Greek market.

8.3.9   Diminishing the Importance and Influence of Multilateral Regimes Such as the EU

This indicator also emerged with respect to the Burgas-Alexandroupolis pipeline project in the 1990s and 2000s. The Russian Federation did not negotiate with the European Union as a whole, instead preferring to negotiate with individual countries. Negotiations with Greece and Bulgaria took place on a bilateral and/or trilateral basis.

8.3.10   Attempts to Control the Entire Supply Chain (Regardless of Commercial Rationale)

No such effort was found in Greece. Actions connected to the Burgas-Alexandroupolis pipeline project and the Hellenic Petroleum S.A. purchase plan in 2001–02 rather reflected a broader regional strategy, one not exclusively targeting Greece.

8.3.11   Economically Irrational Steps Taken to Maintain a Particular Position in the Client State's Market

No such effort was found in Greece. Actions connected to the Burgas-Alexandroupolis pipeline project and the Hellenic Petroleum S.A. purchase plan in 2001–02 rather reflected a broader regional strategy, one not exclusively targeting Greece (Table 8.4).

Table 8.4   Summary of indicators

Indicator	Found	Found with
Active support by Russian state representatives for energy enterprises and their activities abroad	Yes	Pipeline Consortium Burgas-Alexandroupolis Ltd. (OAO AK Transneft, 33.34%; OAO NK Rosneft, 33.33%; PAO Gazprom Neft, 33.33%), PAO Lukoil
As a foreign supplier, Russia rewards certain behaviours and links energy deals to the client state's foreign policy orientation	No	–
Abuse of infrastructure (e.g. pipelines) and differential pricing to exert pressure on the client state	No	–
Efforts to take control of the energy resources, transit routes, and distribution networks of the client state	Yes	Pipeline Consortium Burgas-Alexandroupolis Ltd. (OAO AK Transneft, 33.34%; OAO NK Rosneft, 33.33%; PAO Gazprom Neft, 33.33%), PAO Lukoil
Disruption (by various means) of alternative supply routes/sources of supply	No	–
Efforts to gain a dominant market position in the client country	Yes	PAO Lukoil
Efforts to eliminate competitive suppliers	No	–
Acting against liberalization	No	–
Diminishing the importance and influence of multilateral regimes such as the EU	Yes	–
Attempts to control the entire supply chain (regardless of commercial rationale)	No	–
Economically irrational steps taken to maintain a particular position in the client state's market	No	–

Source: Author

Note: *Inconclusive* means some indications of this behaviour were found, but not of a shape, size, or importance to be ascribed to strategic behaviour and thus fulfil the indicator

Sources

Baev, P. K. (2015, July 8). Greece's Russian Fantasy; Russia's European Delusion. *The Brookings blog.* Retrieved from http://www.brookings.edu/blogs/order-from-chaos/posts/2015/07/08-greeces-russian-inspiration-baev

Danchev, S., & Maniatis, G. (2014). *The Refining Sector in Greece: Contribution to the Economy and Prospects.* Athens: Foundation for Economic and Industrial Research. Retrieved from http://iobe.gr/docs/research/en/RES_05_C_27062014_REP_EN.pdf

Energean Oil & Gas S.A. (n.d.). Retrieved from http://www.energean.com/

Eni SpA. (2015). *World Oil and Gas Review 2015.* Rome: Eni SpA. Retrieved from http://www.eni.com/en_IT/attachments/azienda/cultura-energia/wogr/2015/WOGR-2015-unico.pdf

Eqbali, A. (2016, January 25). Greece to Buy 25% of Crude Oil Imports from Iran Under Hellenic Supply Deal. *Platts McGraw Hill Financial.* Retrieved from http://www.platts.com/latest-news/oil/tehran/greece-to-buy-25-of-crude-oil-imports-from-iran-26347122

Hegedüs, K. (2010). Russia's Relations: The Turkish-Greek-Cypriot Triangle. *International Relations Quarterly 1*(2, Summer), Retrieved from http://www.southeast-europe.org/pdf/02/DKE_02_A_W_Hegedus-Krisztina.pdf

Hellenic Petroleum S.A. (n.d.). Retrieved from http://www.helpe.gr/

Iran, Greece Seal Oil Sale Deal. (2016, March 16). *MEHR News Agency.* Retrieved from http://en.mehrnews.com/news/115266/Iran-Greece-seal-oil-sale-deal

International Energy Agency. (2010). *Oil & Gas Security, Emergency Responses of IEA Countries, Greece.* Paris: IEA Publications. Retrieved from https://www.iea.org/publications/freepublications/publication/greece_2010.pdf

International Energy Agency. (2011). *Energy Policies of IEA Countries, Greece, 2011 Review.* Paris: IEA Publications. Retrieved from https://www.iea.org/publications/freepublications/publication/Greece2011_unsecured.pdf

International Energy Agency. (2015). *Energy Supply Security 2014, Emergency Responses of IEA Countries.* Paris: IEA Publications. Retrieved from https://www.iea.org/publications/freepublications/publication/ENERGYSUPPLYSECURITY2014.pdf

Jirušek, M., Vlček, T., & Henderson, J. (2017). Russia's Energy Relations in Southeastern Europe: An Analysis of Motives in Bulgaria and Greece. *Post-Soviet Affairs, 33*(5), 335–355. https://doi.org/10.1080/1060586X.2017.1341256

Kalinkov, K. (2010). Regional Aspects of Burgas-Alexandroupolis Petrol Pipeline. *Analele ştiinţifice ale Universităţii Alexandru Ioan Cuza din Iaşi, 57*(2010), 63–73. Retrieved from http://anale.feaa.uaic.ro/anale/resurse/ec1k-alinkov.pdf

Leonard, M., & Popescu, N. (2007). *A Power Audit of EU-Russia Relations.* European Council on Foreign Relations policy paper. Retrieved from http://fride.org/uploads/file/A_power_audit_of_relations_eu-russia.pdf

Littler, A., Pitsas, N., & Jørgensen, V. B. (2015). *Statement on Security of Energy Supply Republic of Macedonia.* Strengthening the Administrative Capacity of the Energy Department in the Ministry of Economy and the Energy Agency, EuropeAid/129822/D/SER/MK. Skopje: ATC Consultants GmbH and Its

Consortium Partners. Retrieved from https://www.energy-community.org/portal/page/portal/ENC_HOME/DOCS/3844261/21A784C4C0A96A7 5E053C92FA8C0392A.PDF

Lukoil wycofuje się z Polski. (2016, February 5). *TVN24BiS.pl.* Retrieved from http://tvn24bis.pl/surowce,78/lukoil-wycofuje-sie-z-polski,616849.html

Mearns, E. (2015, July 10). Oil Imports Have Energy Poor Greece In A Stranglehold. *OilPrice.com.* Retrieved from http://oilprice.com/Energy/Energy-General/Oil-Imports-Have-Energy-Poor-Greece-In-A-Stranglehold.html

Ministry of Development of the Hellenic Republic. (2007). *"Burgas–Alexandroupolis" Oil Pipeline.* Retrieved from http://www.minpress.gr/minpress/en/enhmerwtiko_entypo_mpoyrgas.pdf

Michaletos, I. (2011, March 20). The Greek Energy Sector in 2011: Corporate Profiles of the Major Players. *Balkanalysis.com.* Retrieved from http://www.balkanalysis.com/greece/2011/03/20/the-greek-energy-sector-in-2011-corporate-profiles-of-the-major-players/

Motor Oil Hellas S.A. (n.d.). Retrieved from http://www.moh.gr/

New Greek Government: Russia's Trojan Horse Inside the EU? (2015, January 28). *Intellinews.com.* Retrieved from http://www.intellinews.com/new-greek-government-russia-s-trojan-horse-inside-the-eu-500443388/?source=russia& archive=bne

OKTA Crude Oil Refinery A.D. (2014). *Annual Report, Financial Statements and Annual Accounts For the year ended 31st December 2013.* Skopje: OKTA Crude Oil Refinery A.D. Retrieved from http://www.okta-elpe.com/uploads/PDF/FS%2031%20December%202013.pdf

Olerex to Purchase Lukoil Service Stations in Estonia. (2015, June 7). *NewEuropeInvestor.com.* Retrieved from http://www.neweuropeinvestor.com/news/lukoil-olerex-estonia-10381/

Onti, N. M. (2013, March 7). Greece Still Relies On Russia For Oil. *Greek Reporter.* Retrieved from http://greece.greekreporter.com/2013/03/07/greece-still-relies-on-russia-for-oil/

Papaspanos, J. (2010). *Caspian Energy Geopolitics: The Rise and Fall of Burgas-Alexandroupoli.* Research Paper No. 148 of the Research Institute for European and American Studies, Athens. Retrieved from http://www.rieas.gr/images/rieas148.pdf

Prometheus Gas S.A. (n.d.). Retrieved from http://www.prometheusgas.gr/

Poussenkova, N. (2012). "They Went East, They Went West…": The Global Expansion of Russian Oil Companies. In P. Alto (Ed.), *Russia's Energy Policies, National, Interregional and Global Levels* (pp. 185–205). Cheltenham and Northampton: Edward Elgar Publishing Limited.

Putin and Purvanov in Athens: The Long Road to Burgas-Alexandroupolis. (2006, September 5). *WikiLeaks*. Retrieved from https://wikileaks.org/plusd/cables/06ATHENS2324_a.html

Russia's Lukoil Selling Assets in Lithuania, Latvia Due to Local Anti-Russia Sentiment. (2015, December 24). *TASS Russian News Agency*. Retrieved from http://tass.ru/en/economy/846725

Sideris, S. (2015, June 24). Novak: Bulgaria Is Not Interested in the 'Revival' of the Burgas-Alexandroupolis Project. *Independent Balkan News Agency*. Retrieved from http://www.balkaneu.com/novak-bulgaria-interested-revival-burgas-alexandroupolis-project/

Time for Mergers and Acquisitions in the Oil and Gas Industry in Europe. (2015). *Petroleum Industry Review 2015* (April). Retrieved from http://www.petroleumreview.ro/magazine/2015/april-2015/45-april-2015/552-time-for-mergers-and-acquisitions-in-the-oil-and-gas-industry-in-europe

Trans-Balkan Pipeline B.V. (n.d.). Retrieved from http://www.tbpipeline.com/

Tsachevsky, V. (2011). *Bulgaria, the Balkans and the Pan-European Infrastructure Projects*. Electronic Publications of Pan-European Institute, 1/2011. Turku: Turku School of Economics. Retrieved from https://www.utu.fi/fi/yksikot/tse/yksikot/PEI/raportit-ja-tietopaketit/Documents/Tsachevsky_netti_final_2011.pdf

Tziampiris, A. (2010). Greek Foreign Policy and Russia: Political Realignment, Civilizational Aspects, and Realism. *Mediterranean Quarterly, 21*(2), 78–89. https://doi.org/10.1215/10474552-2010-006

Wattles, J. (2015, July 12). Russia May Throw Greece an Energy 'Lifeline'. *CNN Money*. Retrieved from http://money.cnn.com/2015/07/12/news/economy/russia-oil-gas-greece/

List of Interviews

Interview 01: Athens, Greece, April 18, 2016.
Interview 02: Athens, Greece, April 18, 2016.
Interview 03: Athens, Greece, April 18, 2016.
Interview 04: Athens, Greece, April 19, 2016.
Interview 05: Athens, Greece, April 19, 2016.
Interview 06: Athens, Greece, April 19, 2016.

CHAPTER 9

Republic of Macedonia

9.1 Crude Oil Sector General Information

9.1.1 Introduction and Upstream

The Republic of Macedonia is a small, landlocked Central Balkan country with approximately two million inhabitants, neighbouring Albania, Kosovo, Serbia, Bulgaria, and Greece. Small in size, the country consumed little crude oil in the past. In 2013, crude oil was completely supplanted by oil product imports, when OKTA Crude Oil Refinery A.D., the only refinery in the republic, ceased operation. Macedonian crude oil imports were formerly in the range of 900,000 tonnes per year in the 1990s and 2000s; 259,000 tonnes were imported in 2012 (Republic of Macedonia State Statistical Office, 2013), but as noted, no crude oil has been imported at all since 2013. Imported petroleum products made up 34.2% of gross inland consumption and 36.4% of final energy consumption in industry in 2014 (Republic of Macedonia State Statistical Office, 2015, p. 12). The most recent decade shows a stable division of petroleum product consumption among the three major sectors: transport (50–55%), industry (25–30%), and heat and electricity production (5–10%) (Republic of Macedonia State Statistical Office, n.d.).

There are no crude oil reserves in the Republic of Macedonia and no extraction takes place. A few companies from Norway, Turkey, Kazakhstan, and Canada expressed interest in prospecting for oil in 2006. The American

© The Author(s) 2019
T. Vlček, M. Jirušek, *Russian Oil Enterprises in Europe*,
https://doi.org/10.1007/978-3-030-19839-8_9

company Schlumberger Limited did some work on oil surveys in Macedonia (Interview 03), but their outcome failed to persuade any of the companies that had shown interest to do business in the country.

9.1.2   Midstream

There is but a single pipeline in the country, running 213 km from the Thessaloniki oil terminal in Greece to the OKTA Crude Oil Refinery A.D. in Skopje. The pipeline, known as Vardax, more or less follows the river Vardar and has a capacity of 2.5 million tonnes annually (mta) (Petroleum Development Consultant Limited & Energetski Institut Hrvoje Požar, 2011, p. 80). It is owned and operated by Crude Oil Pipeline Company Thessaloniki-Skopje-Vardax S.A., which is itself owned by EL.P.ET. Balkaniki S.A.[1] (80%) and the Government of the Republic of Macedonia (20%) (Ministry of Economy of the Republic of Macedonia, 2010, p. 40).

The OKTA refinery, however, closed as indicated earlier in January 2013, and it was agreed that it would thenceforth serve as the main oil product distribution hub. This also impacted the pipeline, which was now to be repurposed for the transport of oil products (Littler, Pitsas, & Jørgensen, 2015, pp. 71–72). Development, however, has been slow: the pipeline was cleaned in 2013–15, but the Macedonian government has so far declined to permit its use for the distribution of oil products ('*Криза со горивата*', 2017). Oil products are thus imported mostly by road transport and by rail (Petroleum Development Consultant Limited & Energetski Institut Hrvoje Požar, 2011, p. 81); imports come mainly from Greece and Bulgaria.

9.1.3   Downstream

The OKTA Crude Oil Refinery A.D., which since 1999 has been owned by EL.P.ET. Balkaniki S.A. (81.51%), Pucko Petrol (10.87%), and Company employees (7.62%) (Hellenic Petroleum S.A., n.d.), operated the now defunct OKTA refinery in Skopje. The refinery launched operations in 1982. Even though it was designed with a capacity of 2.5 mta, the maximum it achieved was 1.36 mta in 1988. The waiving of custom duties

---

[1] EL.P.ET. Balkaniki S.A. is a company jointly controlled by Hellenic Petroleum S.A. (63%) and Consortium of banks APE (37%) (OKTA Crude Oil Refinery A.D., 2014, p. 13).

Table 9.1 Capacity of Macedonian refinery as of 2015

Refinery	Owner	Refining capacity	Type	Year established
OKTA	OKTA Crude Oil Refinery A.D. (81.51% EL.P.ET. Balkaniki S.A.; 10.87% Pucko Petrol; 7.62% Company employees)	2.5 mta	Hydroskimming	1980

Source: Author

for petroleum products since 2011 has rendered the refinery uncompetitive in the Balkan region, and petroleum product imports are now cheaper than domestic production (Ministry of Economy of the Republic of Macedonia, 2010, p. 39, 127). Any relaunch of the refinery is extremely improbable given its age and the technology employed (Interview 02; Interview 03) (Table 9.1).

The downstream market is dominated by three companies: OKTA A.D., Makpetrol A.D., and Lukoil Macedonia DOOEL Skopje; all three participate in the wholesale market. In addition to its former refining operations, OKTA A.D. is active in wholesale and distribution, with 470,000 m³ of oil storage capacity in OKTA installations. The company runs a small network of 36 petrol stations (14% of the market). Makpetrol A.D. operates the largest network of petrol stations in the Republic of Macedonia, with 124 stations (48% of the market), and holds another 75,000 m³ in oil storage capacity. Makpetrol A.D.'s major shareholder[2] is the Republic of Macedonia (approximately 50%; Interview 02; Interview 03), followed by the Macedonian fuel transporter OILKO KDA (22.267%), itself a subsidiary of Makpetrol A.D. This cluster of companies was apparently created for the illegal transfer of revenues from Makpetrol A.D. to OILKO KDA. The president of the Executive board of Makpetrol A.D., Andreja Josifovski, and other managers and board members have been charged with defrauding the operation of EUR 6.1 million by

[2] In the 2000s, interest to purchase the company was shown by the Russian-Israeli-British company Balkan Petroleum Holding Limited with very shady businesspeople and oligarchs Uri Bider (Isreali), Vasili Evdokimov (British), Mikhail Chorny (Russian), and Serhiy Kurchenko (Ukrainian) behind it. Later in 2016, two Russian millionaires, Alexander Kaplan and Alexander Smuzikov (ex-TNK-BP employees), also expressed interest in buying out Makpetrol A.D. As it currently stands, nothing has been realized (Casule & Zhdannikov, 2016; 'Кои се луѓето', 2016).

transferring 19.95% of the shares of Makpetrol A.D. to OILKO KDA, which was formed in 2006. Andreja Josifovski, the Chief Executive Officer (CEO) and largest private shareholder in both companies, is the key suspect as the business decision to invest in OILKO KDA was never approved by the Assembly of Makpetrol shareholders ('*CEO of Makpetrol*', 2015; '*Makpetrol's Executive*', 2015). And yet the whole story seems to have simply faded from public view as no information is available on further developments in the case. Lukoil Macedonia DOOEL Skopje (a subsidiary of Russia's PAO Lukoil) runs a network of 28 petrol stations (11% of the market). Approximately a quarter of the market is held by an assortment of small companies (Littler et al., 2015, p. 73; Lukoil Macedonia DOOEL Skopje, n.d.; Makpetrol A.D., n.d.; МАКПЕТРОЛ АД, 2015, p. 29; Ministry of Economy of the Republic of Macedonia, 2010, p. 40).

9.2   Recent Market Developments and Russian Activities

The Republic of Macedonia, situated as it is in the middle of the Balkan Peninsula, has recently become a strategic focus in discussions regarding planned and proposed pipelines that would run through the Peninsula. This is especially true for natural gas pipelines, but the planned AMBO oil pipeline[3] supports this argument as well. The Republic of Macedonia has truly never gotten strategic attention from Russia in the past, and the current development of relations between the two countries is very likely tied to Macedonia's geopolitical position with regard to the pipelines noted. Relations between the two countries are good, chiefly based at present on economic cooperation, tourism, sport, and the support of education through scholarships.[4] Macedonia also refused to support EU and US sanctions against Russia, and Macedonian President Gjorge Ivanov participated in celebrations of the 70th anniversary of Victory Day in Moscow in May 2015. In March 2016, a Macedonian delegation met the Deputy Head of the Russian Federal Tourism Agency (Rosturizm), Sergei Korneev, motivated by the country's strong interest in expanding cooperation with Russia in the tourism sector ('*Македония хочет*', 2016).

---

[3] See the chapter on Bulgaria for more information about the AMBO pipeline.

[4] Numerous young Macedonians study in Russia (one year for the language, then five years at university), with their studies fully covered by the Russian state scholarship programme (Interview 03).

Since the civil unrest in Skopje in 2015, Russia has also become acutely sensitive to developments in the country, a primary goal being to steer the country away from membership in Western institutions (EU, NATO). Opposition to the rule of Prime Minister Nikola Gruevski, whose ten-year reign has been seen as authoritarian, with traces of the governing style of Vladimir Putin—an object of fascination to the pro-government media and analysts (Georgievski, 2015)—has been characterized by Russian Foreign Minister Sergey Lavrov as 'fairly brutally and managed by outsiders' (*'Bulgaria Rejects'*, 2015). Gruevski has been strongly supported by the Russian government. Putin stated in May 2015 that 'Russia supports Macedonia's leaders and their efforts to normalize the political situation in the country. Our interest is a stable and prosperous Macedonia' (*'Putin: The Stability'*, 2015). The crisis had been mediated by the European Union and led to early general elections in December 2016 after Gruevski's resignation from office in January 2016 and interim replacement by Emil Dimitriev, the party treasurer. In the general elections, Zoran Zaev then came to power as the new prime minister of Macedonia. It should be added that there was no sign of involvement by the Russian Federation in these political developments. The political issue at stake was internal and centred on evidence of massive wiretapping and monitoring of citizens made public by the opposition. Gruevski's resignation, the call for early elections, and the composition of the new government were, however, strongly influenced by the EU and the USA (Interview 03).

Russia's primary goal of avoiding Macedonia's entry into the Western institutions was clearly visible in the summer of 2018, when Greece expelled two Russian diplomats who tried to thwart settlement of the Greece-Macedonia issue over Macedonia's name (*'Řecko vyhostí'*, 2018). Russia fears the likely agreement (with the country renamed the Republic of North Macedonia) will make Macedonia acceptable to Western structures—something which has until now been blocked by the naming dispute. Macedonia has been a candidate to join the European Union since 2005 and has also applied for NATO membership. The invitation to NATO was blocked by Greece at the 2008 Bucharest summit.

Still, the relationship between Macedonia and Russia must be understood on a broader basis. The pro-Russian or anti-Western element in Macedonia may be identified with the pro-Serbian stance taken primarily by (ethnic) Macedonian political parties and individuals (Interview 02; Interview 03; Interview 04). The pro-Russian element is therefore an indirect one, in actuality part of pro-Serbian politics and policy. Every

Macedonian government save the current one has been pro-Serbian. The government of Prime Minister Zaev is more West-leaning, but what this really boils down to is that it is pro-Albanian; ethnic Albanian and Muslim political parties and inhabitants support a Western trajectory (Interview 02; Interview 03; Interview 04).

Overall, the Russian Federation is not an important or interesting area of foreign policy for Macedonia. Macedonians have no specific positive ties with Russians; they even lack a common church (Interview 01; Interview 03; Interview 04). Their most important foreign partner is Serbia, even though the only support given is usually oral and lacks any tangible benefits. For Serbia, however, good relations with Macedonia are vital. As a landlocked country, Serbia needs a friendly neighbour to transport a variety of goods across its territory to Serbia. The failure of negotiations to lease the ports of Shengjin in Albania and Neum in Bosnia and Herzegovina left Serbia with Macedonia as its most viable import route for goods from the Greek maritime port of Thessaloniki (some goods are also imported through the port of Dürres, Albania) (Interview 03).

In addition, Russia's PAO Lukoil has made efforts to enhance its position in the Republic of Macedonia, because the country's business climate is very attractive for PAO Lukoil. It entered the Macedonian market in June 2005 after signing a Memorandum of Collaboration with the Macedonian government. Under the agreement, signed by Prime Minister Vlado Bučkovski and PAO Lukoil's CEO Vagit Alekperov in St. Petersburg, the Government of the Republic of Macedonia has undertaken to allocate sites for the construction of 40 PAO Lukoil petrol stations. In 2006, PAO Lukoil purchased the Štip Petrol Storehouse, an oil storage facility with a capacity of 5600 m^3 and opened its first petrol station in Skopje. Since that time, the company has opened a network of 28 petrol stations, falling short of its goal of 40 stations by 2010 ('*Lukoil Macedonia Wins*', 2012; '*LUKoil to Invest*', 2005; '*LUKoil to Open*', 2006).

The relationship between the Republic of Macedonia and PAO Lukoil is warm, with high-ranking officials meeting on a regular basis and statements of further cooperation and investment plans being issued often. Recent meetings have seen Prime Minister Gruevski meet with PAO Lukoil CEO Vagit Alekperov in June 2012; also at that time, Nikola Gruevski met with Vladimir Putin; and in October 2013 the president of the Republic of Macedonia, Gjorge Ivanov, met PAO Lukoil Vice President Vadim Vorobyev. The June 2012 meeting gave strong impetus to the relations between these two entities, with PAO Lukoil announcing increased investment in the republic and Gruevski stating that Macedonia would 'become a regional centre of Lukoil, also covering Bulgaria, Albania and Kosovo' (Government of Republic of Macedonia, 2012; 'LUKOIL Plans', 2012). In the 2013 meeting, Vorobyev announced that PAO Lukoil would 'continue developing its retail network in Macedonia to reach 20% market share' ('The Vice-President', 2013). Even the recent withdrawal of PAO Lukoil from many Central and Eastern Europe (CEE) countries due to the unfavourable business environment[5] has not affected its position in the Republic of Macedonia. This supports the argument that very good relations exist between the two countries on both the official and public levels.

These efforts follow quite logically on PAO Lukoil's acquisition of the Neftochim refinery in Bulgaria in 1999 and the Petrotel refinery in Romania in 1998. The company has been working on its position either by building oil retail networks in the individual countries of the Balkan Peninsula or by closing supply contracts with other retail companies to sell products from these refineries. The development of the petrol station network in the Republic of Macedonia is thus logically integrated with its market operations in the region.

9.3   Research Indicator Assessment

9.3.1   Active Support by Russian State Representatives for Energy Enterprises and Their Activities Abroad

This indicator was positive several times, but most meetings with Russian officials, especially with Vladimir Putin, focused on the natural gas sector. Whether there has been Kremlin support for PAO Lukoil's actions in the Republic of Macedonia is debatable—it is the company's representatives who take part in meetings with Macedonian politicians. Instead, Kremlin support has been linked to the cultivation of a pro-Russian (pro-Serbian) political environment in Macedonia in general, as was the case support for pro-Russian Prime Minister Nikola Gruevski.

9.3.2   As a Foreign Supplier, Russia Rewards Certain Behaviours and Links Energy Deals to the Client State's Foreign Policy Orientation

No strict connection was found between this indicator and Macedonia's crude oil sector.

---

[5] See the chapter on Greece for more information.

9.3.3   Abuse of Infrastructure (e.g. Pipelines) and Differential Pricing to Exert Pressure on the Client State

Since 2013 there has been no functioning crude oil pipeline for import in the Republic of Macedonia, and the country does not in fact import crude oil at all. All petroleum product consumption depends upon imports. The majority of these products are of non-Russian origin (Greece, Croatia, and other EU countries) with some coming from the PAO Lukoil refinery in Bulgaria (a small portion of the imports of Makpetrol A.D. and all the imports of Lukoil Macedonia DOOEL Skopje). Lukoil Macedonia DOOEL Skopje's petrol stations represent 11% of the Macedonian market.

Even though the country is 100% dependent on petroleum product imports, the absolute numbers are small (805,000 tonnes in 2014; Littler et al., 2015, p. 73). The existence of multiple retail and wholesale companies, the fact that Lukoil Macedonia DOOEL Skopje holds only 11% of the market, and the option of importing oil products from several locations via highway transport make it structurally difficult to exert influence over infrastructural bottlenecks.

9.3.4   Efforts to Take Control of the Energy Resources, Transit Routes, and Distribution Networks of the Client State

Lukoil Macedonia DOOEL Skopje has not engaged in any effort to seize control of distribution networks or transit routes. No extraordinary measures or activities by the company (chosen by PAO Lukoil for the best company outside Russia in a corporate competition in 2012) were uncovered.

9.3.5   Disruption (by Various Means) of Alternative Supply Routes/Sources of Supply

No such effort has been indicated.

9.3.6   Efforts to Gain a Dominant Market Position in the Client Country

Even though Lukoil Macedonia DOOEL Skopje's strategy is to expand to a point that it controls 20% of the market (equal to approximately 40 petrol stations), this can hardly be labelled as an effort to gain a dominant market position. Makpetrol A.D. operates the biggest network of petrol stations in the Republic of Macedonia, and it has 124 petrol stations (48% of the market) and possesses 75,000 m³ of oil storage capacity. Its position is clearly dominant and no indication that PAO Lukoil intends to acquire the company or merge it into Lukoil Macedonia DOOEL Skopje was found.

There is no overarching Russian strategy in the country, but some businessmen of Russian nationality (in tandem with others) have not been shy about attempting to bribe and corrupt local politicians and businessmen to secure a better market position in the country, to avoid public procurement procedures, or to get to state contracts (Interview 02).

9.3.7   Efforts to Eliminate Competitive Suppliers

No indication of efforts on the part of PAO Lukoil to eliminate competition, such as acquiring competing companies or merging them into Lukoil Macedonia DOOEL Skopje, was found.

9.3.8   Acting Against Liberalization

As stated in the 2015 Energy Community report, '[t]he oil market in Macedonia is primarily driven by market-based forces, although the maximum refining and retail prices for oil derivatives and the maximum retail prices for blends of fossil fuels and biofuels are set pursuant to the price-setting regulations that are issued by the Energy Regulatory Commission' (Littler et al., 2015, p. 71). No effort against liberalization was discovered.

9.3.9   Diminishing the Importance and Influence of Multilateral Regimes Such as the EU

No such effort has been indicated.

9.3.10   Attempts to Control the Entire Supply Chain (Regardless of Commercial Rationale)

No such effort has been indicated.

9.3.11   Economically Irrational Steps Taken to Maintain a Particular Position in the Client State's Market

No such steps were found to have been taken in the Republic of Macedonia (Table 9.2).

Table 9.2   Summary of indicators

Indicator	Found	Found with
Active support by Russian state representatives for energy enterprises and their activities abroad	Inconclusive	PAO Lukoil
As a foreign supplier, Russia rewards certain behaviours and links energy deals to the client state's foreign policy orientation	No	–
Abuse of infrastructure (e.g. pipelines) and differential pricing to exert pressure on the client state	No	–
Efforts to take control of the energy resources, transit routes and distribution networks of the client state	No	–
Disruption (by various means) of alternative supply routes/sources of supply	No	–
Efforts to gain a dominant market position in the client country	No	–
Efforts to eliminate competitive suppliers	No	–
Acting against liberalization	No	–
Diminishing the importance and influence of multilateral regimes such as the EU	No	–
Attempts to control the entire supply chain (regardless of commercial rationale)	No	–
Economically irrational steps taken to maintain a particular position in the client state's market	No	–

Source: Author

Note: *Inconclusive* means some indications of this behaviour were found, but not of a shape, size, or importance to be ascribed to strategic behaviour and thus fulfil the indicator

Sources

Bulgaria Rejects Russian Claim that It Wants to Dismember Macedonia. (2015, May 21). *EurActiv.com*. Retrieved from http://www.euractiv.com/section/global-europe/news/bulgaria-rejects-russian-claim-that-it-wants-to-dismember-macedonia/

Casule, K., & Zhdannikov, D. (2016, November 2). Russian Investors Stalled in Bid for Macedonia's Makpetrol. *Reuters*. Retrieved from http://uk.reuters.com/article/makpetrol-ma-investors/russian-investors-stalled-in-bid-for-macedonias-makpetrol-idUKL8N1D34T2

CEO of Makpetrol Charged with Fraud. (2015, August 14). *MINA – Macedonian International News Agency*. Retrieved from http://macedoniaonline.eu/content/view/27923/45/

Georgievski, B. (2015, May 26). Macedonia: A Pawn in the Russian Geopolitical Game? *Deutsche Welle*. Retrieved from http://dw.com/p/1FWSD

Government of Republic of Macedonia. (2012, June 22). *PM Gruevski: Macedonia to Become 'Lukoil' Regional Center.* Retrieved from http://www.vlada.mk/node/3616?language=en-gb

Hellenic Petroleum S.A. (n.d.). Retrieved from http://www.helpe.gr/

Кои се луѓето што се обидуваат да ја преземат Макпетрол? (2016, June 3). *Нова ТВ*. Retrieved from http://novatv.mk/koi-se-lugeto-shto-se-obiduvaat-da-ja-prezemat-makpetrol/

Криза со горивата има оти владата две години не дозволува нафтоводот да работи, обвини Хеленик Петролеум. (2017, February 8). *САКАМДАКАЖАМ. МК*. Retrieved from http://sdk.mk/index.php/makedonija/kriza-gorivata-ima-oti-vladata-dve-godini-ne-dozvoluva-naftovodot-da-raboti-obvini-helenik-petroleum/

Littler, A., Pitsas, N., & Jørgensen, V. B. (2015). *Statement on Security of Energy Supply Republic of Macedonia.* Strengthening the Administrative Capacity of the Energy Department in the Ministry of Economy and the Energy Agency, EuropeAid/129822/D/SER/MK. Skopje: ATC Consultants GmbH and Its Consortium Partners. Retrieved from https://www.energy-community.org/portal/page/portal/ENC_HOME/DOCS/3844261/21A784C4C0A96A75E053C92FA8C0392A.PDF

Lukoil Macedonia DOOEL Skopje. (n.d.). Retrieved from http://www.lukoil.com.mk/

Lukoil Macedonia Wins Best Foreign Unit Award. (2012, July 29). *MINA – Macedonian International News Agency.* Retrieved from http://macedoniaonline.eu/content/view/21555/1/

LUKOIL Plans to Increase the Investment in Macedonia. (2012, June 21). *Lukoil Macedonia DOOEL Skopje.* Retrieved from http://www.lukoil.com.mk/en/news/LUKOIL-investment-Alekeprov

LUKoil to Invest in FYROM Project. (2005, June 19). *NewEurope.eu.* Retrieved from https://www.neweurope.eu/article/lukoil-invest-fyrom-project/

LUKoil to Open 40 Petrol Stations in Macedonia in Four Years. (2006, September 8). *Sputnik News.* Retrieved from http://sputniknews.com/business/20060908/53665218.html

Македония хочет расширить сотрудничество с Россией в сфере туризма. (2016, March 25). *РИА Новости.* Retrieved from http://ria.ru/tourism/20160325/1397158283.html

Makpetrol A.D. (n.d.). Retrieved from http://www.makpetrol.com.mk/

МАКПЕТРОЛ АД. (2015). *XIX СЕДНИЦА НА СОБРАНИЕ НА АКЦИОНЕРИ.* Retrieved from https://www.makpetrol.com.mk/Download/Sobranie%202015/Izvestaj%20za%20SOBRANIE%20MP%202015.pdf

Makpetrol's Executive and 20 Other Affiliates Accused in a Financial Scheme Worth 6.1 million Euro. (2015, August 13). *Новинска агенција "Мета.мк".* Retrieved from http://meta.mk/en/prviot-chovek-na-makpetrol-i-ushte-20-litsa-osomnicheni-za-kriminal-tezhok-nad-6-milioni-evra/

Ministry of Economy of the Republic of Macedonia. (2010). *Strategy for Energy Development in the Republic of Macedonia Until 2030.* Retrieved from http://www.ea.gov.mk/projects/unece/docs/legislation/Macedonian_Energy_Strategy_until_2030_adopted.pdf

OKTA Crude Oil Refinery A.D. (2014). *Annual Report, Financial Statements and Annual Accounts For the Year ended 31st December 2013.* Skopje: OKTA Crude Oil Refinery A.D. Retrieved from http://www.okta-elpe.com/uploads/PDF/FS%2031%20December%202013.pdf

Petroleum Development Consultant Limited, & Energetski Institut Hrvoje Požar. (2011). *Emergency Oil Stocks in the Energy Community Level.* Retrieved from https://www.energy-community.org/portal/page/portal/ENC_HOME/DOCS/2516177/0633975AB2907B9CE053C92FA8C06338.PDF

Putin: The Stability of Macedonia Is in Russia's Interest. (2015, May 28). *Новинска агенција "Мета.мк".* Retrieved from http://meta.mk/en/putin-stabilnosta-na-makedonija-e-interes-na-rusija/

Řecko vyhostí ruské diplomaty, snažili se ovlivnit dohodu s Makedonií. (2018, July 11). *ECHO24.cz.* Retrieved from https://echo24.cz/a/SggEb/recko-vyhosti-ruske-diplomaty-snazili-se-ovlivnit-dohodu-s-makedonii

Republic of Macedonia State Statistical Office. (2013, November 15). *Balance of Oil by Months, 2012.* Retrieved from http://www.stat.gov.mk/pdf/2013/6.1.13.82.pdf

Republic of Macedonia State Statistical Office. (2015, October 20). *Energy Balances, 2014.* Retrieved from http://www.stat.gov.mk/pdf/2015/6.1.15.78.pdf

Republic of Macedonia State Statistical Office. (n.d.). Retrieved from http://www.stat.gov.mk/

The Vice-President of OAD LUKOIL, Mr. V.N. Vorobyev Visit the Republic of Macedonia. (2013, October 24). *Lukoil Macedonia DOOEL Skopje.* Retrieved from http://www.lukoil.com.mk/en/news/Vorobjov

LIST OF INTERVIEWS

Interview 01: Ohrid, Macedonia, August 29, 2017.
Interview 02: Ohrid, Macedonia, August 29, 2017.
Interview 03: Skopje, Macedonia, August 30, 2017.
Interview 04: Skopje, Macedonia, August 30, 2017.

CHAPTER 10

Romania

10.1   CRUDE OIL SECTOR GENERAL INFORMATION

10.1.1   Introduction and Upstream

Romania, which considers itself a Danube Region country and not part of South-East Europe or the Balkans, is a large country of nearly 240,000 km² with 19.5 million inhabitants. It borders Ukraine, Moldova, Bulgaria, Serbia, and Hungary, and is washed by the Black Sea to the East. Crude oil makes up 34.4% of Romania's total primary energy supply (TPES; 2014) and petroleum products account for 32% of total final consumption (2014). The vast majority of petroleum products are consumed in the transport sector (68.9% in 2014) (International Energy Agency Statistics, n.d.). Romania's oil consumption has hovered around 9 million tonnes annually (mta) over the last decade. Only around 55% of the country's crude oil needs are imported, as Romania has an extensive tradition of oil extraction. Some original oil production operations are still active today. By comparing the information in Table 10.1, which states that around 55% of Romania's refinery intake is imported, with KMG International N.V.'s data stating that approximately 40% of Romania's imports come from Kazakhstan (KMG International, 2016, p. 18), we arrive at a maximum of 15% of Romania's needs that are imported from Russia (ca. 1.35 mta).

The National Agency for Mineral Resources has successfully encouraged foreign companies to invest in the exploration and production of

© The Author(s) 2019   151
T. Vlček, M. Jirušek, *Russian Oil Enterprises in Europe*,
https://doi.org/10.1007/978-3-030-19839-8_10

Table 10.1 Romania crude oil data

	2009	2012	2013	2014
Consumption	9.19	9.21	8.40	8.95
Production	4.30	3.91	3.90	3.81
Import dependency (%)	53.2	57.5	53.6	57.4

Source: Oil Consumption, 2015; OMV Petrom S.A., 2011, p. 11, 2015, p. 9

Note: Figures in millions of tonnes; percentage calculations by T. Vlček

hydrocarbons. Since the fall of communism, ten public auctions have been held for both onshore and offshore oil reserves in Romania. A number of companies have undertaken oil exploration in Romania, yet the vast majority of domestic production (98.5% in 2014) is produced by OMV Petrom S.A., which operates 232 commercial oil and gas fields in the country (OMV Petrom S.A., 2016, p. 24). OMV Petrom S.A. collaborates with the Hunt Oil Company (Texas, USA), Exxonmobil Exploration and Production Romania Ltd. (a subsidiary of Exxon Mobil Corporation), and Repsol S.A. (Spain). Among the numerous companies investing in exploration, S.C. Stratum Energy Romania LLC (a 100% subsidiary of the USA's Stratum Energy Group Ltd.) had already started oil and gas recovery in Bacău County by 2014.

Ranked as it is among the top five European countries for proven crude oil reserves, with 0.5 billion barrels (68 million tonnes), Romania has generated interest from oil companies (BP, 2016, p. 6). But lacking significant investment to locate new deposits in the country, current reserves will last only about 17 years at the current rate of exploration. For this reason, Romanian Minister of Energy Victor Grigorescu is trying to attract companies to invest in oil exploration as part of a new strategy that involves opening new oil fields for tender concessions (the last was the tenth such tender, which took place in 2010). Unlike with natural gas, offshore deposits in the Black Sea will probably not yield a major discovery, so exploration efforts are focused instead on onshore locations. Today, onshore exploration accounts for 96% of oil production in Romania, while the deposits in the Black Sea are unfortunately promising only for natural gas ('*Romania Vying*', 2016).

10.1.2   *Midstream*

Crude oil deliveries to Romanian refineries take place via the 1540 km domestic pipeline network that connects oil fields with refineries in Ploiești, Steaua Română, Dărmănești, and Onești, and the 1348 km network of pipelines for imported crude oil that connects the Oil Terminal in Constanța with refineries in Pitești, Ploiești, and Onești. The system, with a total length of 2888 km, is called the Crude Oil National Transport System (Sistemul Național de Transport al petrolului, S.N.T.) and is owned by S.C. CONPET S.A. The company's shareholders are the Romanian Ministry of Energy, Small and Medium Enterprises (58.7%), and other legal (33.5%) and private (7.8%) entities (S.C. CONPET S.A., 2016, p. 9). The company is administered by the National Agency for Mineral Resources (Agenția Națională pentru Resurse Minerale, n.d.).

Since 2012, the volumes of imported oil have varied between 1.5 and 2.9 mta, even though the capacity of the pipelines is much greater, at 20.2 mta (S.C. CONPET S.A., 2016, p. 5, n.d.). At the moment, the system essentially supplies only OMV Petrom S.A.'s and S.C. Petrotel-Lukoil S.A.'s refineries in Ploiești. This information is supported by the fact that these two companies were the major business partners of S.C. CONPET S.A. in the first half of 2015, accounting for 98.1% of turnover (S.C. CONPET S.A., n.d.) (Table 10.2).

The small transport volumes of imported crude oil stem from two factors. The first is that refineries make partial use of domestic oil as feedstock for their operations, and the second is that the 5 mta Petromidia refinery ceased to use the pipeline in 2008 when its owner, KMG International

Table 10.2 Volumes of transported crude oil via the Crude Oil National Transport System

	2012	2013	2014	2015
Domestic crude (mta)	4.302	3.929	3.932	3.838
Utilization rate (%)[a]	62.3	56.9	57.0	55.6
Imported crude (mta)	1.488	1.806	2.473	2.907
Utilization rate (%)[a]	7.4	8.9	12.2	14.4
Total	5.790	5.735	6.405	6.745

Source: S.C. CONPET S.A., 2016, p. 5, n.d.

Note: Utilization rate calculated by T. Vlček

[a]The capacity of pipelines for domestic crude oil is 6.9 mta, and 20.2 mta for imported crude oil

N.V., constructed the Midia Oil Terminal to supply the refinery with oil from Kazakhstan (see later).

Between 2002 and 2015, a pipeline project connecting the Eastern and Western shores of the Balkan Peninsula was proposed and discussed. The project, called Pan-European Oil Pipeline (PEOP), was intended to begin at the Romanian port of Constanţa and stretch through Romania, Serbia, Croatia, and Slovenia to the Italian port of Trieste. In 2008, a project company headquartered in London was created, named Pan-European Oil Pipeline Project Development Company (PEOP-PDC). Its shareholders were Jadranski naftovod d.d. (33.3%), JP Transnafta (33.3%), S.C. CONPET S.A. (16.65%), and Oil Terminal SA Constanţa (16.65%) (Romania Ministry of Economy, 2009, p. 2). The pipeline, which was to be 1319 km in length with a planned capacity of 40–90 mta, was meant to supply oil not only to the refineries along its South-East European route (in Romania, Serbia, and Croatia), but also to Hungary and Slovakia via a connection to the Adria pipeline system, and to Germany and the Czech Republic via a connection to the TAL pipeline starting in Trieste. The Romanian government, as the majority shareholder in S.C. CONPET S.A., saw a major business opportunity with high revenues coming from the transit fees. To attract regional attention, the project was presented as one of the best options for overcoming environmental risks as well as the limited capacity of the Turkish Straits (Mihajlovic, 2010, p. 5). But Italy's reluctance to support the project, Slovenia's environmental issues because of plans to run the route through the Karst Plateau, and Croatia's eventual withdrawal in 2010 meant that the project failed to attract interest and investors. The president of the Administrative Board of Transnafta, Bratislav Čeperković, announced in February 2010 that the first phase of the project—the connection between Constanţa and Pančevo—was still alive ('*Romania and Serbia*', 2010). The three companies—JP Transnafta, S.C. CONPET S.A., and Oil Terminal SA Constanţa—thus continued with the project; however, in June 2014, the Minister Delegate for Energy of Romania, Răzvan Nicolescu, announced that 'the project no longer has any chance of success' ('*Minister Delegate*', 2014). In September 2014, the procedures to liquidate PEOP-PDC started ('*Project Development*', 2014).

10.1.3   Downstream

The extensive tradition of the Romanian petroleum industry is clearly visible in the fact that there are 11 refineries in the country. Six, however, are either on hold or closed, and two do not refine crude oil. The remaining three thus constitute the Romanian refining sector as it is today and involve its key players. The total capacity of Romania's crude oil refineries stands at 11.9 mta, and this capacity is controlled by three players.

Rompetrol Rafinare S.A. operates the Petromidia (5 mta) refinery, giving it a 42% share of active refining capacity. Rompetrol Rafinare S.A.'s major shareholders are the Romanian Ministry of Energy, Small and Medium Enterprises and Business Environment (44.7%), and KMG International N.V. (48.1%). KMG International N.V. is the new name (since 2014) of The Rompetrol Group N.V., 75% owned by KazMunayGas Refining and Marketing JSC, a 100% subsidiary of Kazakh state-owned AO NK KazMunayGas.

The second player is OMV Petrom S.A., a 51% subsidiary of Austria's OMV AG (with a 33.2% stake in Romania's hands via two state-owned entities), operating the 4.5 mta Petrobrazi refinery, thereby possessing 37.8% of the active refining capacity in Romania.

The third player is S.C. Petrotel-Lukoil S.A., a 100% subsidiary of PAO Lukoil, with the 2.4 mta Petrotel refinery controlling 20.2% of active refining capacity in Romania (Table 10.3).

Table 10.3 Capacity of Romanian refineries and oil production establishments as of 2017

Refinery	Owner	Design refining capacity (mta)	Nelson complexity index (2015)	Year established
Arpechim (Pitești)	OMV Petrom S.A.	6.5	Closed in 2011	1964
Astra (Ploiești)	Rafinaria Astra Romana S.A.	0.9	On hold since 2004; in insolvency since 2014	1880
Crișana (Suplacu de Barcău)	SC Ecodiesel S.r.l.	0.48	Closed during 2005–2010; closed again in 2011 and for sale since then	1969
Dărmănești	Rafinăria Dărmănești S.A.[a]	0.6	On hold since 2003; closed in 2015	1949
Lubrifin (Brașov)[b]	Lubrifin S.A.	0.067	–	1879

(*continued*)

Table 10.3   (continued)

Refinery	Owner	Design refining capacity (mta)	Nelson complexity index (2015)	Year established
Petrobrazi (Ploieşti)	OMV Petrom S.A.	4.5	11.1	1934
Petromidia (Năvodari)	Rompetrol Rafinare S.A.[c]	5	10.2	1975
Petrotel (Ploieşti)	S.C. Petrotel-Lukoil S.A.[d]	2.4	12.2	1904
Rafo Oneşti	Rafo S.A.	2.6–3	On hold since 2008	1969
Steaua Română	S.C. Rafinaria Steaua Română S.A. Câmpina[e]	0.5	On hold since 2009; in insolvency since 2013	1897
Vega (Ploieşti)	Rompetrol Rafinare S.A.[c]	0.305	–	1906

Source: Compiled by T. Vlček from public sources

[a]Refinery owned by Rafo S.A., which is owned by Austria's Petrochemical Holding GmbH
[b]A lubricant refinery that does not refine crude oil but uses oil feedstocks from local markets, typically supplied by the refineries Steaua Română, Astra, and Petrotel
[c]Rompetrol Rafinare S.A.'s major shareholders are the Romanian Ministry of Energy, Small and Medium Enterprises and Business Environment (44.7%), and KMG International N.V. (48.1%). KMG International N.V. is the new name (since 2014) of the Rompetrol Group N.V., 75% owned by KazMunayGas Refining and Marketing JSC, a 100% subsidiary of Kazakh state-owned AO NK KazMunayGas. Rompetrol Rafinare S.A.'s Vega refinery uses semiproducts from the Petromidia refinery as its feedstock
[d]A 100% subsidiary of PAO Lukoil
[e]82.46% owned by S.C. Omnimpex Chemicals, S.R.L.

The Petromidia refinery is conveniently located on the Black Sea coast near Constanţa; this position allows for deliveries of crude oil using tanker ships to the Midia Oil Terminal (Terminalul petrolier Midia). The Terminal was constructed in 2008 and is also operated by SC Midia Marine Terminal S.R.L., a subsidiary of KMG International N.V. EURO 5 category diesel and gasoline are the main products of the Petromidia refinery; semiproducts from the refinery are used as feedstock for the Vega refinery, which is also part of KMG International N.V. The vast majority of oil comes from Kazakhstan.

OMV Petrom S.A.'s Petrobrazi refinery accepts crude oil from both domestic production (via domestic pipelines) and via the pipeline network of S.C. CONPET S.A. It focuses on producing EURO 5 fuels for its major Romanian petrol station network.

S.C. Petrotel-Lukoil S.A.'s refinery in Petrotel refines both crude oil from domestic production (transported by rail) and Russian Export Blend from Ural imported by the pipeline network of CONPET S.A. from the Oil Terminal in Constanţa (PAO Lukoil, 2014, p. 43, n.d.). It produces mainly medium distillates such as gas oil, sulphur, and coke (50% in 2014), and gasoline (32% in 2014) (PAO Lukoil, n.d.).

There are approximately 2000 petrol stations in Romania. The main retail players are OMV Petrom S.A., S.C. Lukoil Romania S.R.L., KMG International N.V. (The Rompetrol Group N.V.), and MOL Romania Petroleum Products S.R.L. OMV Petrom S.A. operates 554 petrol stations (28% of the retail market), of which 153 petrol stations offer the premium OMV brand. Its main competitor, KMG International N.V. (The Rompetrol Group N.V.), has a network of 405 petrol stations and additionally controls 155 Partner stations and 126 Express stations. This makes for a 34% share in control of the retail market in Romania. S.C. Lukoil Romania S.R.L. operates a network of 307 petrol stations (15% of the retail market) and MOL Romania Petroleum Products S.R.L. has been on the Romanian market since 1994, purchasing the networks of the Amoco Corporation (1997), Shell Romania S.R.L. (2004), and Eni Romania S.R.L. (2015). Today the company has a network of 200 petrol stations (10% of the market). The remaining approximately 250 petrol stations are operated by smaller companies, such as SOCAR Petroleum S.A. (35 stations), S.C. Oscar Downstream S.R.L., and NIS Petrol s.r.l. (18 stations; a 100% subsidiary of Naftna industrija Srbije a.d., which is 56.15% owned by PJSC Gazprom Neft).

10.2   RECENT MARKET DEVELOPMENTS AND RUSSIAN ACTIVITY

Russian-Romanian relations were cut off abruptly at the beginning of the 1990s, and it took several years to create a new basis for cooperation. Its foundation was the 2003 Treaty on Friendly Relations and Cooperation between Romania and the Russian Federation, which opened a dialogue between the two countries and allowed for economic and cultural cooperation. The relationship was later strengthened by the signing of the Agreement on Economic Cooperation between the Government of Russia and the Government of Romania in 2007 and the meeting of the presidents of the two countries, Vladimir Putin and Traian Băsescu, at the NATO summit in Bucharest in 2008 (Feyt, 2014, p. 54; '*President Vladimir*', 2008). These events initiated a decade of fairly intensive communication between the two countries, especially in the areas of economy and energy. Around 20 meetings of representatives from different Romanian state institutions[1] took place with their Russian counterparts in Bucharest, Moscow, and elsewhere starting in 2008 (Embassy of Romania in the Russian Federation, n.d.). Energy was discussed on a regular basis with meetings of representatives from OAO Gazprom (Alexander Medvedev, Alexei Miller) and other companies.

At present, Russian-Romanian relations may be characterized by a common interest in enhancing mutual economic ties and cooperation contrasted with disagreements that centre on the historical region of Bessarabia, that is, the Republic of Moldova and the unrecognized, self-proclaimed Pridnestrovian Moldavian Republic.[2] During the Crimean crisis, Romania was one of the most strident voices calling for harsh sanctions against Russia, and this provoked deterioration in Russian-Romanian relations. With Crimea now under Russian control, Russia has become a close neighbour (across the Black Sea) and there is a risk that it will challenge Romania's ownership rights to oil and gas deposits in the Black Sea, or act to influence exploration and exploitation activity.

Russian companies have been present within the oil sector since 1998. In 1997–98 PAO Lukoil was successful in the privatization process of the Petrotel refinery. A 51% share of Petrotel S.A. was sold to PAO Lukoil for US $52 million (International Business Publications, 2016, p. 76). The company also promised to invest US $189 million to modernize the refinery, another US $11 million for environmental purposes, and promised to take over the refinery's US $53 million debt (Vatansever, 2006, p. 13).

Having purchased the refinery, PAO Lukoil established a network of subsidiaries in Romania, including S.C. Petrotel-Lukoil S.A. (refining), S.C. Lukoil Romania S.R.L. (retail), S.C. LUKOIL Energy & Gas Romania S.R.L. (production and sale of electricity), Lukoil Lubricants East Europe S.R.L. (production and sale of lubricants), Lukoil Technology Services Romania S.R.L. (research, design, development, implementation, and maintenance services in IT and automation), and a Lukoil Overseas office in Bucharest. Over the years, PAO Lukoil has stabilized its position, holding 20.2% of the active refining capacity in Romania and 15% of the

---

[1] The Chamber of Commerce and Industry; Minister of the Economy, Commerce, and Business Environment; Ministry of Small and Medium Enterprises, Commerce, and Business Environment; Minister of Agriculture and Rural Development; Ministry of Regional Development and Tourism.

[2] See Rodkiewicz, 2011, for more detailed information.

retail market via its network of 307 petrol stations. However, no plans or efforts to dominate the Romanian market are evident, nor is there any interest in purchasing other capacities or assets.

S.C. Petrotel-Lukoil S.A. has been recently accused of avoiding taxes due to the Romanian state, and prosecutors, police, and customs seized oil and fuel inventories worth EUR 230 million in October 2014 (*'Romania President Fuels'*, 2014). This spurred a tense exchange between the two sides, starting with PAO Lukoil Vice President for Refining, Marketing, and Distribution Vladimir Nekrasov, who intimated that PAO Lukoil might permanently close the refinery in reaction to the seizure of assets (*'Romanian President: Government'*, 2014), with the loss of 3500 employee jobs. Romanian President Traian Băsescu reacted by declaring the pressure as unacceptable and stating that the Romanian government would be ready to buy out PAO Lukoil's share in the refinery (*'President Basescu'*, 2014; *'Romania President Fuels'*, 2014). In July 2015, Kajel Holdings Limited (a 100% subsidiary of China Peace Petroleum Group) showed interest in buying the refinery, but a spokesman for PAO Lukoil stated the company had no plans to sell the refinery (*'Russia's Lukoil'*, 2015). In December 2015, S.C. Petrotel-Lukoil S.A. signed a new contract worth EUR 9.7 million with S.C. CONPET S.A. to transport crude oil to the Petrotel refinery (*'Romanian Oil'*, 2015). This clearly confirmed PAO Lukoil's continued interest in the Petrotel refinery and showed that the brief closure of the refinery in October 2014 simply related to the investigation.

OAO TNK-BP and OAO Gazprom had expressed interest in the Romanian market earlier, in 2002–04, when they took part in the second round of the Petrom S.A. privatization. OAO Gazprom bid for the tender with six other competitors, with Austria's OMV AG chosen as the winner (Pitersky, 2004). OAO Gazprom is thus currently present in the Romanian oil market strictly via NIS Petrol s.r.l. (a 100% subsidiary of Naftna industrija Srbije a.d. 56.15% owned by PJSC Gazprom Neft) with 18 petrol stations, less than 1% of the retail market.

10.3   Research Indicator Assessment

No Russian efforts related to the research indicators were found in Romania. With reference to the indicator '[a]ctive support by the Kremlin for its state-owned energy enterprises and their activities in a given country', support was aimed at the natural gas business and not present for oil.

Certain elements were found that would appear to fall under the indicator '[a]cting against liberalization in energy sectors and diversification'. In 2008, six companies in the Romanian fuel market entered into a cartel agreement to jointly pull one kind of gasoline off the market. These companies included OMV Petrom S.A., S.C. Lukoil Romania S.R.L., Eni Romania S.R.L., The Rompetrol Group N.V. (today KMG International N.V.), and MOL Romania Petroleum Products S.R.L. The cases involved were heard by Romania's High Court; the case involving Lukoil Romania S.R.L. is still being tried; the other companies have lost to the Romanian Competition Council and must pay fines ('*Romanian High*', 2016; '*Romania's Competition*', 2016). But this situation must be understood for what it is—nothing more than an illegal attempt to maximize profits.

SOURCES

Agenția Națională pentru Resurse Minerale. (n.d.). Retrieved from http://www.namr.ro/

BP. (2016). *BP Statistical Review of World Energy June 2016*. London: BP p.l.c. Retrieved from https://www.bp.com/content/dam/bp/pdf/energy-economics/statistical-review-2016/bp-statistical-review-of-world-energy-2016-full-report.pdf

Embassy of Romania in the Russian Federation. (n.d.). *Economic Relations Between Romania and Russia Federation*. Retrieved from https://moscova.mae.ro/en/node/395

Feyt, N. (2014). Russian-Romanian Relations in the 21st Century. *Political Science and International Relations, XI*(2), 53–59. Retrieved from http://journal.ispri.ro/wp-content/uploads/2014/07/53-59-Nadezda.pdf

International Business Publications. (2016). *Romania Company Laws and Regulations Handbook Volume 1 Strategic Information and Basic Laws*. Washington: International Business Publications.

International Energy Agency Statistics. (n.d.). Retrieved from https://www.iea.org/statistics/

KMG International. (2016). *Economic Impact Assessment*. Bucharest: KMG International / Bucharest University of Economic Studies. Retrieved from http://www.rompetrol.com/sites/default/files/kmgi_study_en_preview.pdf

Mihajlovic, S. (2010, November 8–9). *Presentation of Transnafta at the 2nd Oil Forum of the Energy Community*. Belgrade, Serbia. Retrieved from https://www.energy-community.org/portal/page/portal/ENC_HOME/DOCS/772192/0633975AAFD57B9CE053C92FA8C06338.PDF

Minister Delegate for Energy: 'Constanta-Trieste Project Has No Longer Any Chance of Success'. (2014, June 7). *The Diplomat Bucharest*. Retrieved from http://www.thediplomat.ro/articol.php?id=5149

Oil Consumption – Romania. (2015, December 31). *Quandl*. Retrieved from https://www.quandl.com/data/BP/OIL_CONSUM_ROU-Oil-Consumption-Romania

OMV Petrom S.A. (2011). *OMV Petrom Group in Figures 2010*. Bucharest: OMV Petrom S.A. Retrieved from https://www.omvpetrom.com/SecurityServlet/secure?cid=1255733817458&lang=en&swa_site=&swa_nav=&swa_pid=&swa_lang=

OMV Petrom S.A. (2015). *OMV Petrom Group in figures 2014*. Bucharest: OMV Petrom S.A. Retrieved from https://www.omvpetrom.com/SecurityServlet/secure?cid=1255764915303&lang=en&swa_site=&swa_nav=&swa_pid=&swa_lang=

OMV Petrom S.A. (2016). *Annual Report 2015*. Bucharest: OMV Petrom S.A. Retrieved from https://www.omvpetrom.com/SecurityServlet/secure?cid=1255770325643&lang=en&swa_site=&swa_nav=&swa_pid=&swa_lang=

PAO Lukoil. (2014). *Fact Book 2014*. Retrieved from http://www.lukoil.com/materials/doc/IR2014/FB43.pdf

PAO Lukoil. (n.d.). Retrieved from http://www.lukoil.com/

Pitersky, A. (2004). The Key to Success. *Oil of Russia 2004* (2). Retrieved from http://www.oilru.com/or/16/187/

President Basescu Slams Lukoil, Asks Gov't to Get Ready to Take Over Petrotel. (2014, October 9). *The Romania Journal*. Retrieved from http://www.romaniajournal.ro/president-basescu-slams-lukoil-asks-govt-to-get-ready-to-take-over-petrotel/

President Vladimir Putin Met with Romanian President Traian Basescu. (2008, April 4). *President of Russia Events*. Retrieved from http://en.kremlin.ru/events/president/news/44077

Project Development Company for PEOP To Be Liquidated. (2014, September 31). *Petroleum Industry Review*. Retrieved from http://www.petroleumreview.ro/magazine/2015/dec-jan-2015/31-september-2014/254-project-development-company-for-peop-to-be-liquidated

Rodkiewicz, W. (Ed.). (2011). *Transnistrian Conflict After 20 Years A Report by an International Expert Group*. Warsaw and Chisinau: Centre for Eastern Studies/Institute for Development and Social Initiatives 'Viitorul'. Retrieved from https://www.osw.waw.pl/sites/default/files/transnistrian_conflict_after_20_years.pdf

Romania and Serbia Continue Realization of Pan-European Oil Pipeline. (2010, February 12). *ekapija Poslovni Portal*. Retrieved from http://www.ekapija.com/website/en/company/photoArticle.php?id=293202&path=naftovod26.jpg

Romania Ministry of Economy. (2009). *Presentation Sheet of the Investment Project Pan European Oil Pipeline (PEOP) Constanta – Trieste Pipeline.* Bucharest: Romania Ministry of Economy. Retrieved from http://www.minind.ro/invest/new/Oil_and_Gas_Sector/CONPET/OG.C.1.09_peop_Engl.pdf

Romania President Fuels Row Over Russian Refinery. (2014, October 9). *Balkan Insight.* Retrieved from http://www.balkaninsight.com/en/article/romania-president-adds-presure-on-lukoil-refinery-dispute

Romania Vying for New Investments in Oil and Gas. (2016, March 4). *Business Review.* Retrieved from http://www.business-review.eu/sidebar-featured/romania-vying-for-new-investments-in-oil-and-gas-invest-in-romania-exclusive-report-100063

Romania's Competition Council to Enforce Record Cartel Fine on Oil Firms – Final Decision. (2016, April 22). *Business Review.* Retrieved from http://www.business-review.eu/news/romanias-competition-council-to-enforce-record-cartel-fine-on-oil-firms-final-decision-104348

Romanian High Court Trims Fuel Cartel Fines By $24.2M. (2016, April 25). *Law360.* Retrieved from http://www.law360.com/articles/788453/romanian-high-court-trims-fuel-cartel-fines-by-24-2m

Romanian Oil Pipeline Operator Conpet Signs 9.7 mln Euro Deal with Petrotel Lukoil Refinery. (2015, December 30). *SeeNews.* Retrieved from https://seenews.com/news/romanian-oil-pipeline-operator-conpet-signs-97-mln-euro-deal-with-petrotel-lukoil-refinery-456085

Romanian President: Government Should Be Prepared to Take Over Lukoil Refinery if Russians Decide to Shut It Down. (2014, October 9). *Romania Insider.* Retrieved from http://www.romania-insider.com/romanian-president-government-should-be-prepared-to-take-over-lukoil-refinery-if-russians-decide-to-shut-it-down/

Russia's Lukoil Approached by Chinese Firm for Romanian Refinery: Letter. (2015, July 20). *Reuters.* Retrieved from http://www.reuters.com/article/us-romania-lukoil-idUSKCN0PU1UI20150720

S.C. CONPET S.A. (2016). *Annual Report of the Administrators of the Company 'CONPET' S.A. For the Financial Year Ended December 31st 2015.* Ploieşti: S.C. CONPET S.A. Retrieved from http://www.conpet.ro/wp-content/uploads/2016/04/1.-Administrators-audited-annual-report-2015.pdf

S.C. CONPET S.A. (n.d.). Retrieved from http://www.conpet.ro/

Vatansever, A. (2006). *Russian Involvement in Eastern Europe's Petroleum Industry, The Case of Bulgaria.* London: GMB Publishing Ltd.

CHAPTER 11

Republic of Serbia

11.1   Crude Oil Sector General Information

11.1.1   Introduction and Upstream

The Republic of Serbia is one of the largest Balkan countries with 10.5 million inhabitants spread over an area of approximately 77,000 km². Serbia is truly a Central Balkan country, with eight countries as neighbours and no access to the sea. Crude oil and natural gas liquids make up only 11% of primary energy production in Serbia, as coal is primarily used to meet the country's energy needs. Nevertheless, petroleum refining products, which are chiefly utilized in the transport sector, account for 33.7% of Serbia's total energy consumption (Statistical Office of the Republic of Serbia, 2016a, p. 27, 2016b, p. 61). The country's gross crude oil consumption has ranged between 2.1 and 2.9 million tonnes annually over the most recent five years. Petroleum products and crude oil are major import items and approximately half of all energy imports are petroleum products and crude oil. These imports have stabilized recently at a level of approximately 1.6 million tonnes of crude oil annually and 1 million tonnes of petroleum products annually. Until 2010, all imported crude oil officially came through Naftna industrija Srbije a.d. based on the guaranteed monopoly status of NIS a.d. (see below). Even though the monopoly was revoked in January 2011, all crude oil is still being imported by NIS a.d., since it is the only refiner in Serbia. Since 2008, all oil has been purchased through public tenders. Oil is purchased from many intermediaries

© The Author(s) 2019
T. Vlček, M. Jirušek, *Russian Oil Enterprises in Europe*,
https://doi.org/10.1007/978-3-030-19839-8_11

Table 11.1   Republic of Serbia crude oil data

	2009	2012	2013	2014
Consumption	2.716	2.203	2.826	2.742
Production	0.666	1.130	1.234	1.165
Import dependency	75.5%	48.7%	56.3%	57.5%

Source: Energy Community, n.d.

Note: Figures in thousands of tonnes; percentage calculations by T. Vlček

(in general non-Russian), but more than 95% of crude oil originates in the Russian Federation (REB) or Kazakhstan (CPC), with some small amounts coming from Iraq (Kirkuk, Basra) (Interview 05) (Table 11.1).

Serbia's extensive history of oil production dates back to the Yugoslav era, with the first oil field having been discovered in 1952. The main Serbian oil fields are Velebit, Kikinda, Mokrin, Rusanda, Elemir, Kikinda-Varoš, and Turija Sever. All oil production in the country is done by Naftna industrija Srbije a.d. (NIS a.d.), with 666 wells currently under exploitation in 42 fields (Naftna industrija Srbije a.d., n.d.). Since 2009, NIS a.d. has been majority-owned by PJSC Gazprom Neft, a subsidiary of the Russian OAO Gazprom. There are also many oil shale deposits in Serbia, with major deposits in Aleksinac, Senonian Trench, Babušnica, and Niš. The Aleksinac deposit is the most promising and most developed. Serbia is currently seeking an advisor to prepare a tender for selecting a strategic partner for oil shale exploration in Aleksinac. Oliver Dulić, Serbian Minister of Environment, Mining and Spatial Planning, said 'oil shale could be exploited for several decades, with an annual production of between 500,000 and 600,000 tons of crude, 100 megawatts of electricity, and enough thermal energy to heat the city of Aleksinac and neighbouring villages' (Zimonjic, 2012). All oil produced in Serbia can be refined in Serbian refineries. Most domestic production is light sweet oil, but the Serbian refineries are able to refine heavier fractions, too (Interview 03) (Table 11.2).

Production has been growing since 2010 (see Table 11.1) following the NIS a.d. 2011–2020 Strategy of Geological Research. The document envisages further expansion of oil production in the country until 2020. All Serbian crude oil is refined in domestic refineries. The company also invests in exploration, especially in Southern Serbia, in the Vojvodina region, in the Pozarevac Depression, and in Central Serbia. Major reserves

Table 11.2 Shareholders structure of Naftna industrija Srbije a.d.

Shareholder	Share (%)
PJSC Gazprom Neft	56.15
Republic of Serbia	29.87
Societe Generale Bank Srbija a.d.—Custodian Account	0.50
Unicredit Bank Srbija a.d.—Custodian Account	0.34
Raiffeisen bank a.d. Beograd—Custodian Account	0.20
Aktiv-Fond d.o.o.	0.14
AWLL Communications d.o.o. Beograd	0.14
Unicredit Bank Srbija a.d.—Joint Account	0.14
Global Macro Capital Opportunities	0.13
Keramika Jovanovic d.o.o. Zrenjanin	0.12
Other shareholders	12.25

Source: Naftna industrija Srbije a.d., n.d.

Note: Data as of June 2016

have recently been located in these localities. Currently 50 million tonnes of oil reserves have been confirmed, with assessments by various experts ranging up to 170 million tonnes (*'All Serbian Oil'*, 2015).

11.1.2 *Midstream*

Crude oil is mostly transported by pipeline, with the exception of domestic crude from the oil fields of Turija, Janošik, and Stig, which is carried by tanker trailers (Ministry of Energy, Development and Environmental Protection of the Republic of Serbia, 2013, p. 52).

The pipeline route that supplies crude to Serbian refineries begins at the oil terminal of Omišalj, a port on the Croatian island Krk. The stretch that runs through Croatia is part of a larger system called Jadranski Naftovod (JANAF).[1]

In 1999, after NATO's bombing of former Yugoslavia, NIS a.d. was granted a monopoly until the end of 2010 by the Serbian government on crude oil and oil derivatives imports and on the processing of oil derivatives to allow the company time to recover and become competitive (Dąborowski, 2011; *'Serbia, Russia'*, 2008). The reasoning behind this was twofold: first to recover from the bombing by making NIS a.d. the sole importer of crude oil (importing oil products was banned, meaning

---

[1] See detailed information about the JANAF system in the chapter on Croatia.

consumers had to buy products refined by the monopolist NIS a.d. and thereby contribute to revenues and competitiveness). The paramount aim was to use the revenue to modernize the refinery so that it could produce products that met euro quality standards (Interview 02; Interview 05; Interview 06). The second reason was to combat the grey economy (until 2001, all oil products imports disappeared into the grey economy) by controlling imports and thus price (Interview 05).

The pipeline section from the Croatian-Serbian border city of Sotin to Novi Sad is 63.3 km long, and the subsequent section from Novi Sad to Pančevo is 91 km long (Ministry of Energy, Development and Environmental Protection of the Republic of Serbia, 2013, p. 55). The entire pipeline is owned and operated by JP Transnafta, founded in 2005 and wholly owned by the Government of the Republic of Serbia. The company's average transport volume is 3 mta, despite pipeline capacity of 9 mta between Bosanski Brod and Novi Sad and 6 mta between Novi Sad and Pančevo.

With but a single route to supply oil to Serbia, security is understood to be at stake (Interview 06), and there are some ideas as to how to secure diversification, such as prolonging the Romanian pipeline from Constanţa to the Arpechim Refinery in Piteşti by another 635 km to extend to Pančevo. This is part of the PEOP pipeline project which stretches from Constanţa, Romania to Trieste, Italy.[2] Another option is to use the Danube River, but the river is unreliable and has but limited capacity. The Danube is the longest river in Europe, crossing ten countries and connecting the North Sea via the Rhine to the Black Sea. But in spite of this, operations on the river are at quite a low level—merely 10% of maximum capacity (Radojcic, 2012, p. 1). Seven sections of the river regularly show depths under the 2.5 m necessary to be designated an international waterway. Because of the clogged riverbed, river cargo boats are often loaded only halfway so that they may proceed safely, boosting costs and making for late deliveries (Trejbal, 2012). But the Danube also attains high water levels; it has done so frequently in recent years, and this, too, is a consistent problem.

On the other hand, the diversification projects seem not to be particularly pressing, probably because Serbia has never experienced any curtailment of oil in its modern history and the current route is sufficient in terms of both capacity and security of supply.

---

[2] See the chapter on Romania for more information.

11.1.3 Downstream

There are two crude oil refineries, in Novi Sad and Pančevo, and three small establishments in Beograd, Kruševac, and Obrenovac. The plants in Beograd and Kruševac are of negligible capacity, focused on motor oil and motor liquids production (local brands Galax, and Fam and Fenix, respectively) and use intermediate goods as inputs. The installation in Obrenovac is a small industrial module for bitumen preparation used to asphalt roads, since the owner is a highway construction company (Interview 04). The two crude oil refineries in Novi Sad and Pančevo are owned by Naftna industrija Srbije a.d.[3] and together have 7.3 mta in nameplate refining capacity. Crude oil is currently processed only in Pančevo because the Novi Sad refinery has been on standby due to reconstruction (Table 11.3).

The Pančevo refinery is able to refine both light and heavy fractions. In fact, its product portfolio covers 70% light fractions (such as LPG, diesel, gasoline) and 30% heavy fractions (such as bitumen, oil). This means the

Table 11.3 Capacity of Serbian refineries and oil production establishments

Refinery	Owner	Refining capacity[a]	Nelson complexity index (2014)	Year established
Novi Sad	Naftna industrija Srbije a.d.	2.5/0 mta	2.5	1968
Pančevo	Naftna industrija Srbije a.d.	4.8/2.4 mta	8.9	1959
Beograd	Rafinerija nafte a.d. Beograd[b]	0.9/0.3 mta	–	1934
Kruševac	Fabrika maziva—FAM a.d.	0.28 mta	–	–
Obrenovac	Tim Putevi d.o.o.[c]	–	–	–

Source: Compiled by T. Vlček from public sources

[a]Nameplate capacity/actual capacity; nameplate capacity only for Kruševac
[b]Owned by the Greek company Neochimiki Group
[c]The company Tim Putevi d.o.o. is a roads construction company with a small industrial module for bitumen preparation used to asphalt roads

---

[3]Naftna industrija Srbije a.d. is also a shareholder in HIP-Petrohemija JSC, the largest producer of petrochemicals in the Republic of Serbia. The shareholder structure of HIP-Petrohemija JSC is Republic of Serbia (54.89%), Development Fund of the Republic of Serbia (13.63%), JP Srbijagas (13.38%), Naftna industrija Srbije a.d. (12.72%), PD Elektrovojvodina LLC (4.87%), and City of Pančevo (0.51%) (HIP-Petrohemija JSC, n.d.).

modernized refinery is able to refine diverse types of oil and also attain maximum yields in production, that is, more value from the bottom of the barrel (Interview 05).

The refinery in Novi Sad is undergoing reconstruction and moderniza-tion to be able to process domestic crude. Based on an interview with Kirill Kravchenko, the CEO of NIS a.d., the company plans to refine domestic (Velebit) crude for paraffinic and naphthenic base oils[4] for sub-sequent lubricant production in Pančevo. Capacity is planned at 180,000–500,000 tonnes annually, and combined with the completed modernization of the Pančevo refinery, should allow output of 5 mta of petroleum products to be achieved by the plants in 2020 (Kirillov, 2013; Interview 04; Interview 06). The plan is to expand to other Balkan coun-tries, first to countries where NIS a.d. or PJSC Gazprom Neft already operate through their petrol station networks (Bosnia and Herzegovina, Romania, Bulgaria), and second to new markets that include Hungary and Croatia (Interview 05). Yet the Balkan market is quite saturated, and this strategy might never come to fruition. Completion of the modernization project is therefore not completely certain.

In 2009, Comico Oil d.o.o., a Serbian unit of US-Dutch Comico Overseas N.V., announced its interest in building a refinery in Smederevo (40 km from Belgrade) with a capacity of 5 mta. The company won approval from the city to lease land for construction of the refinery (113 hectares area for US $7.7 million for 99 years). However, the company failed to pay for the lease upfront as agreed, and the city filed a lawsuit against Comico Oil d.o.o. The court decided the company must pay by 4 February 2013, which the company failed to do, prompting termination of the contract ('*Beograd hoće*', 2010; Savic, 2012; '*Serbian Town Scraps*', 2013; '*Serbia's Smederevo*', 2012; '*Who Is*', 2013).

The retail market in Serbia consists of approximately 1440 retail sta-tions. NIS Petrol EOOD, a subsidiary of NIS a.d., is the key player; it operates a network of 330 petrol stations. Since 2012, NIS Petrol EOOD has also operated a network of 16 premium petrol stations in Serbia (96 in the Balkans[5] in total) under the Gazprom Petrol Stations brand (Naftna

---

[4]The idea also arose to use the Novi Sad refinery as a facility for refining intermediate products from Pančevo and to build a product pipeline between the two refineries. But the project was infeasible, and construction was never initiated. No oil products from Pančevo are used for operations at Novi Sad.

[5]Serbia, Bulgaria, Romania, and Bosnia and Herzegovina.

industrija Srbije a.d., n.d.; Ministry of Energy, Development and Environmental Protection of the Republic of Serbia, 2013, p. 56). Taken together, this gives the company control of 24% of the market. Lukoil Srbija a.d. (subsidiary of Russian PAO Lukoil) operates a network of 113 stations (7.8% of the market),[6] OMV Serbia d.o.o. 61 stations (4.2% of the market), EKO Serbia a.d. (owned by Greek Hellenic Petroleum S.A.) 51 petrol stations (3.5% market share), MOL Serbia d.o.o. 33 stations (2.3% of the market), Petrol d.d. 9 stations, and there are numerous other small local companies operating more than 800 retail stations which remain (Ristić, 2011; Souček & Vasiliev, 2014; Statistical Office of the Republic of Serbia, n.d.).[7]

11.2   Recent Market Developments and Russian Activities

The Republic of Serbia and the Russian Federation have enjoyed cooperation and partnership over the long term. But contrary to the popular view, these relations cannot be traced back to the Cold War era. Josip Broz Tito's Yugoslavia (which was never in the USSR) could hardly be called a partner to the USSR. Instead, the beginning of the partnership dates to the 1990s, especially the late 1990s, when it was firmly cemented. It was also in this era that the Russian Federation began to issue strident propaganda based on sentiment, emotion, and image-making, resonating especially powerfully with lower-income and older citizens (Interview 01; Interview 02).

Several incidents occurred during the Yugoslavian wars that brought the two countries into closer alliance; one was the imposition of Western sanctions against Yugoslavia, which forced the country to rely almost exclusively on Russian imports. But the key event was the 1999 NATO aerial bombing of Yugoslavia, which was strongly condemned by Russia and strengthened ties between the two countries considerably. During the NATO bombing, the Serbian parliament even sent a request to Moscow to join the Russian Federation (Interview 02) and the bombing still lives in the nation's memory. In fact, according to public opinion polls in Serbia, 80% of Serbs oppose any cooperation with NATO (Trifkovic, 2016).

---

[6] PAO Lukoil purchased Beopetrol a.d. after a privatization tender in 2003, where it competed against Hungarian MOL Rt. The Privatization Agency of the Government of Serbia sold 79.5% of the company shares for EUR 117 million and an obligation to invest EUR 85 million to its development ('Serbia Suspects', 2013).

[7] There are 463 retail licences in Serbia (Rajal & Petrović, 2016).

The Kosovo issue eventually resulted in a declaration of independence by Kosovo in 2008, which has not been recognized and is strongly opposed by Serbia and Russia. Even though reluctant to support the Russian annexation of Crimea in its initial stages in 2014, Belgrade later provided informal military assistance to pro-Russian separatists in Donbas (Ramani, 2016). Serbia later showed the quality of bilateral relations when it refused to impose sanctions on Russia in 2015, nor has it participated in anti-Russian sanctions in the period since ('*Belgrade Will Not*', 2016; Interview 02).

Officials from the two countries meet on a regular basis in Serbia and Russia, as well as on international platforms. The two countries even participated in joint military exercises in 2014 (Srem) and 2015 (Slavic Brotherhood, together with Belarus). On the other hand, Serbia participates in NATO military exercises on a regular basis and Serbia cannot rely strictly on Russia; indeed, the country has been a candidate for membership in the European Union since 2012 and needs Western partners. Serbian officials thus carefully balance foreign policy, for example, by balancing efforts to enter NATO (Serbia joined NATO's Partnership for Peace programme, signed the Status of Forces Agreement, and joined the NATO Individual Partnership Action Plan) with plans to purchase Russian military equipment (MIG-29 fighter jets and S-300 missile systems), and so on. This balancing act on the part of Serbia is not just a political tactic but may have become a political necessity. Under the treaties noted above, NATO soldiers are permitted free transit through Serbia, may use the Serbian military infrastructure, and are granted immunity (Trifkovic, 2016). Russia would like to have a similar treaty in place with Serbia and clear pressure is being applied: Russian Deputy Prime Minister Dmitry Rogozin told Serbian Foreign Minister Ivica Dačić that signing an agreement with Russia would be in the interests of Serbia's valued military and political neutrality (Reljić, 2016, p. 25).

The true background of Russia's relationship with Serbia may be clearly discerned in the Kosovo issue. It is what prompted Russia to enter the Serbian oil market. The support Russia provided during the Kosovo crisis by vetoing in the Security Council any UN endorsement of a declaration of independence ('*Gazprom's Bid*', 2008; Sekularac, 2014) and by blocking Kosovo's UN membership was 'repaid' by the framework energy agreement signed between the two countries in January 2008. Several plans were agreed as part of the framework, including that Serbia would assist in building the South Stream gas pipeline over its territory (and would acquire the land needed for the pipeline) as well as an underground gas storage facility in Banatski Dvor. It also allowed PJSC Gazprom Neft to acquire a 51% stake in Naftna industrija Srbije a.d. for EUR 400 million in 2008, as well as a 51% share in the local company in charge of both construction and management of the Serbian stretch of the South Stream (Markovic, 2016; 'Serbian Oil and Gas', 2014). PJSC Gazprom Neft also acquired a network of 497 petrol stations, a 43% share in with Serbian daily newspaper Politika, and a number of buildings and hotels in Serbia and Montenegro (including a 38% share in the Hyatt Hotel Belgrade). A pledge to invest EUR 500 million in the refinery complex by 2012[8] was part of the framework agreement negotiated by a team led by Borislav Stefanović, guaranteed by Serbian President Boris Tadić, and signed in Moscow by Petar Škundrić (at that time the Serbian Minister of Energy) and Alexander Dyukov (CEO of PJSC Gazprom Neft). Serbia also agreed to operate under existing law until the implementation of the projects, which meant, for example, that Naftna industrija Srbije a.d. will pay only 3% in mining royalties until 2035 instead of 7% (Markovic, 2016). And it will do so until the company decides the investment in the refinery complex (part of the framework energy agreement) has been repaid, which might be a long time in coming: Naftna industrija Srbije a.d. claims it has been hit hard by sanctions against the Russian oil and gas sector (Gotev, 2016).

Opposition was strong at the time of the sale and continues to be today. Critics decried the cheap sale price in 2008, stating that the true capital asset value of a 51% share in Naftna industrija Srbije a.d. was EUR 993.8 million. Deloitte, meanwhile, assessed the market value of the 51% share at EUR 1.12 billion ('Serbian Oil and Gas', 2014). Mlađan Dinkić, the Minister of Economy at that time, said the government could achieve a price five to eight times the Gazprom bid if it offered the stake via a competitive tender process ('Gazprom's Bid', 2008). All this led in 2014 to the creation of a special investigative team under Interior Minister Nebojša Stefanović to examine the circumstances surrounding the privatization of NIS a.d.; but as of August 2018, the process was still said to be in the preliminary investigation phase. What is more likely is that the investigation never really got under way, and the entire issue will simply fade from the Serbian political scene (Markovic, 2016; Sekularac, 2014; Interview 01).

---

[8] The refinery complex was strongly damaged by NATO bombing in 1999; the investment in repairs and modernization was crucial to maintain NIS a.d. competitiveness.

It should be added that Serbia had tried to privatize Naftna industrija Srbije a.d. even before the financial entry of PJSC Gazprom Neft in 2008. The government hired a consortium consisting of Raiffeisen Investment and Merrill Lynch as consultants for the privatization in 1999. The Councillors suggested privatizing 49% of the company in the first phase. Several companies expressed interest, including PAO Lukoil, MOL Rt, OMV AG, Hellenic Petroleum S.A., Petrol d.d., and BP plc (formerly British Petroleum plc). But the public procurement procedure was unsuccessful as no investor came forth. It was only when the South Stream project was introduced in 2007 that the Russian Federation became interested in making the deal described above for Naftna industrija Srbije a.d. (Interview 05).

Both the current situation and the history of the two countries' relations show minimal benefit for Serbia from the framework energy agreement signed in January 2008. The only benefit Serbia enjoys is Russian support on the Kosovo issue. This leverage for the Russians could, however, disappear in the future. Serbia might eventually decide to fully integrate with the West (NATO and the EU) and ultimately follow this foreign policy course even at the expense of its relations with Russia, which means Serbia would have to agree to recognize Kosovo as a condition of joining the EU. As stated by Dušan Reljić, this Russian instrument of influence would then be worthless (Reljić, 2016, p. 28).

11.3   RESEARCH INDICATOR ASSESSMENT

11.3.1   Active Support by Russian State Representatives for Energy Enterprises and Their Activities Abroad

This indicator was particularly relevant to the framework energy agreement between the two countries signed in January 2008. Serbian Prime Minister Vojislav Koštunica met with Vladimir Putin in 2007 and Serbian Deputy Prime Minister Božidar Đelić met with the Russian Minister of Finance Alexei Kudrin and CEO of OAO Gazprom Alexei Miller in 2008. In addition to these meetings, Serbian and Russian officials meet on a regular basis.

11.3.2   As a Foreign Supplier, Russia Rewards Certain Behaviours and Links Energy Deals to the Client State's Foreign Policy Orientation

This indicator is present with regard to Russian energy projects and interests (privatization of NIS a.d.; South Stream gas pipeline; underground gas storage in Banatski Dvor, etc.) and Russian support over the Kosovo issue. The support Russia provided during the Kosovo crisis by vetoing in the Security Council any UN endorsement of a declaration of independence and blocking Kosovo's membership in the UN is clearly linked to Serbian foreign policy. However, Serbia's foreign policy orientation has not really been penalized by Russia for joining NATO's Partnership for Peace programme, signing the Status of Forces Agreement, or joining the NATO Individual Partnership Action Plan, carefully balancing its foreign policy to reflect its ties to both the West and Russia. This indicator is visible instead outside of pure 'energy' relations, for example, in 2016 when Russian Deputy Prime Minister Dmitry Rogozin told Serbian Foreign Minister Ivica Dačić that signing a military agreement with Russia would be in the interests of maintaining Serbia's valued military and political neutrality (Reljić, 2016, p. 25).

In general, the Russian influence in Serbia is built upon popular sentiment in favour of Russia, on emotion, and on the image of Russia as a friend/partner/brother (Pan-Slavism), as well as on the consanguinity of the autocephalous Orthodox Churches of Russia and Serbia. Russian influence thus mainly targets the Serbian energy sector and the Russian image using nongovernmental organizations, think tanks, associations, student organizations, and relationships with individual politicians and well-known individuals. The effort to polish Russia's image has been enhanced since 2014 by infrastructural projects with a strong positive impact on citizen perceptions of Russia, despite their limited real impact on the Serbian economy (Interview 01). These projects include the rebuilding of a portion of Serbia's railway infrastructure (372 km; OAO Rossijskije schelesnyje dorogi, n.d.) by LLC International Railways (subsidiary of Russian government-owned OAO Rossijskije schelesnyje dorogi) and a current project to rebuild a relatively unimportant section of the Serbian highway system. The railroad project was organized via a public procurement procedure won by a Russian state company, meaning that the Russian export loan of US $940 million ('*Back on Track*', 2016) flows back to the Russian state. It is very likely the highway project will employ the same logic (Interview 01).

11.3.3   Abuse of Infrastructure (e.g. Pipelines) and Differential Pricing to Exert Pressure on the Client State

The entire refining sector is now controlled by PJSC Gazprom Neft (56.15% share in Naftna industrija Srbije a.d.). In 1999, after the NATO aerial bombing of former Yugoslavia, NIS a.d. received a monopoly from the Serbian government running until January 2011 to allow it time to recover and become competitive. The reasoning behind the monopoly was twofold: first to recover from the bombing by making NIS a.d., the sole importer of crude oil (importing oil products was banned, meaning consumers had to buy products refined by the monopolist NIS a.d. and thereby contribute to revenues and competitiveness); and second to combat the grey economy (until 2001, all oil products imports disappeared into the grey economy) by controlling imports and thus price (Interview 05).

With the purchase of 51% of shares in Naftna industrija Srbije a.d. in 2008, the Russian company enjoyed the monopoly noted together with the purchase. However, since 2008, all oil has been purchased via public tenders. Oil is purchased from many different intermediaries (in general non-Russian), but more than 95% of crude oil originates in the Russian Federation (REB) or Kazakhstan (CPC), with some small amounts coming from Iraq (Kirkuk, Basra) (Interview 05). This logic did not change upon expiration of the monopoly in January 2011. The entire development, however, cannot be understood as a Russian abuse of infrastructure, since the Serbian government made the decision well before PJSC Gazprom Neft entered the market and the framework energy agreement between the two countries was signed.

11.3.4   Efforts to Take Control of the Energy Resources, Transit Routes, and Distribution Networks of the Client State

The entire refining sector is now controlled by PJSC Gazprom Neft (56.15% share in Naftna industrija Srbije a.d.). Naftna industrija Srbije a.d. is the only producer of crude oil in Serbia. More than 95% of crude oil originates in the Russian Federation (REB) or Kazakhstan (CPC). As of 2016, there are 40 licences for the wholesale of oil and oil derivatives in Serbia (Rajal & Petrović, 2016). The only crude oil pipeline (from the Croatian-Serbian border city Sotin to Pančevo) is owned and operated by JP Transnafta, founded in 2005 and fully owned by the Government of the Republic of Serbia. No indicators of interest by Russia in taking over the company were registered. PJSC Gazprom Neft thus controls crude oil production and refining in Serbia and is the key player in retail as it controls 24% of the market.

11.3.5   Disruption (by Various Means) of Alternative Supply Routes/Sources of Supply

PJSC Gazprom Neft (56.15% share in Naftna industrija Srbije a.d.) obviously utilizes its ability to refine Russian crude oil. PJSC Gazprom Neft is also in control of the entirety of domestic oil production, thus even these 'alternative' sources are ultimately under its control. The sources of supply are thus under Russian control, while the pipeline route is in the hands of JP Transnafta (wholly owned by Government of the Republic of Serbia).

11.3.6   Efforts to Gain a Dominant Market Position in the Client Country

PJSC Gazprom Neft controls crude oil production and refining in Serbia and is the key player in retail as it controls 24% of the market, making it a dominant market player. NIS a.d.'s development plans up to 2020 envisage refining rising to 5 mta, and sales to 5 mta of petroleum products, which is tied to further prospecting for oil (and gas) domestically as well as outside the borders of Serbia (Hungary, Romania, Bosnia and Herzegovina) (Kirillov, 2013). There are no indicators of interest for even greater expansion (acquisitions, investments in technology, further modernization), but there are signs of interests in getting involved with other sectors, especially electricity generation. This is connected to the construction of a modern 140 MW combined heat and power plant in Pančevo, with the possibility of increasing capacity to 208 MW (Radovanović, 2016, p. 18).

11.3.7   Efforts to Eliminate Competitive Suppliers

No clear evidence was found for this indicator.

11.3.8   Acting Against Liberalization

No clear evidence was found for this indicator.

11.3.9   Diminishing the Importance and Influence of Multilateral Regimes Such as the EU

The privatization of Naftna industrija Srbije a.d. was negotiated and sealed strictly at the bilateral level. No interference from multilateral regimes was registered in the Naftna industrija Srbije a.d. development or elsewhere.

11.3.10   Attempts to Control the Entire Supply Chain (Regardless of Commercial Rationale)

No clear evidence was found for this indicator. The PJSC Gazprom Neft entrance to NIS a.d. was not related to Russia's strategy of controlling the entire supply chain, but agreed to in the framework energy agreement in 2008. In order to have a strategic investor able to modernize and revitalize the Serbian refining sector, Serbia agreed on a low price for the sale of NIS a.d., assistance in building the South Stream gas pipeline through Serbia and the underground gas storage facility in Banatski Dvor, and mining royalties of just 3% instead of 7% for NIS a.d. crude oil extraction operations.

11.3.11   Economically Irrational Steps Taken to Maintain a Particular Position in the Client State's Market

Russian companies' operations in Serbia follow an economic logic, and such steps were not counted. Even the purchase of NIS a.d. in 2008 cannot be taken to be an economically irrational step. The privatization process did not originally attract investors even though several companies had expressed initial interest, including PAO Lukoil, MOL Rt, OMV AG, Hellenic Petroleum S.A., Petrol d.d., and BP plc (formerly British Petroleum plc). It was when the South Stream project was introduced in 2007 that the Russian Federation became interested in making the deal described above for Naftna industrija Srbije a.d. (Interview 05). Yet the target was not to create a commanding position in Serbia's energy sector, but rather to ensure Serbia's assistance in building the South Stream gas pipeline through Serbia and the underground gas storage facility in Banatski Dvor. In short, it was a deal that exchanged Russia's interest in the South Stream for Serbia's interest in revitalizing and modernizing its oil sector (Table 11.4).

Table 11.4 Summary of indicators

Indicator	Found	Found with
Active support by Russian state representatives for energy enterprises and their activities abroad	Yes	See above for details
As a foreign supplier, Russia rewards certain behaviours and links energy deals to the client state's foreign policy orientation	Yes	See above for details
Abuse of infrastructure (e.g. pipelines) and differential pricing to exert pressure on the client state	No	–
Efforts to take control of the energy resources, transit routes, and distribution networks of the client state	No	–
Disruption (by various means) of alternative supply routes/ sources of supply	Inconclusive	–
Efforts to gain a dominant market position in the client country	Inconclusive	–
Efforts to eliminate competitive suppliers	No	–
Acting against liberalization	No	–
Diminishing the importance and influence of multilateral regimes such as the EU	Inconclusive	–
Attempts to control the entire supply chain (regardless of commercial rationale)	No	–
Economically irrational steps taken to maintain a particular position in the client state's market	No	–

Source: Author

Note: *Inconclusive* means some indications of this behaviour were found, but not of a shape, size, or importance to be ascribed to strategic behaviour and thus fulfil the indicator

SOURCES

All Serbian Oil and Gas Reserves Belong to Gazprom. (2015, January 5). *B92*. Retrieved from http://www.b92.net/eng/news/business.php?yyyy=2015&mm=01&dd=05&nav_id=92768

Back on Track: Serbia Looks to Russia for Fresh Upgrade to Railway Network. (2016, March 5). *Sputnik News*. Retrieved from https://sputniknews.com/europe/20160305/1035818315/serbia-russia-train-railways.html

Belgrade Will Not Back the European Union's Sanctions Against Moscow for the Purpose of Mirroring Brussels' Foreign Policy, Serbian President Tomislav Nikolić said During a Meeting with US Vice President Joe Biden. (2016, August 17). *Sputnik News*. Retrieved from https://sputniknews.com/politics/20160817/1044342107/serbia-anti-russia-sanctions.html

Beograd hoće rafineriju nafte. (2010, April 16). *B92*. Retrieved from http://www.b92.net/biz/vesti/srbija.php?yyyy=2010&mm=04&dd=16&nav_id=424893

Dąborowski, T. (2011, 12 January). Russia No Longer Holds a Monopoly on Oil Supplies to Serbia. *OSW Analyses*. Retrieved from http://www.osw.waw.pl/en/publikacje/analyses/2011-01-12/russia-no-longer-holds-a-monopoly-oil-supplies-to-serbia

Energy Community. (n.d.). Retrieved from https://www.energy-community.org/ Gazprom's Bid for Serbia's NIS Brings Pros and Cons for Europe. (2008, January 18). *IHS Markit*. Retrieved from https://www.ihs.com/country-industry-forecasting.html?ID=106597224

Gotev, G. (2016, March 15). Serbia Claims Sanctions Against Russia Hurt Its EU Accession Process. *EurActiv*. Retrieved from https://www.euractiv.com/section/enlargement/news/serbia-claims-sanctions-against-russia-hurt-its-eu-accession-process/

HIP-Petrohemija JSC. (n.d.). Retrieved from http://www.hip-petrohemija.com/

Kirillov, D. (2013, October 30). Interview with CEO, NIS and Deputy CEO, Gazprom Neft, Kirill Kravchenko. *Gazprom Magazine*. Retrieved from http://www.gazprom-neft.com/press-center/lib/1095820/

Markovic, V. (2016, April 25). Privatization of Serbian Oil Company NIS by Gazprom Still in Pre-Investigation Phase. *Mining See*. Retrieved from http://miningsee.eu/privatization-of-serbian-oil-company-nis-by-gazprom-still-in-pre-investigation-phase/

Ministry of Energy, Development and Environmental Protection of the Republic of Serbia. (2013). *Report on the Security of Energy Supply in the Republic of Serbia in the Period from 2011 to 2013*. Belgrade: Ministry of Energy, Development and Environmental Protection of the Republic of Serbia. Retrieved from https://www.energy-community.org/portal/page/portal/ENC_HOME/DOCS/2460177/0633975AD4957B9CE053C92FA8C06338.PDF

Naftna industrija Srbije a.d. (n.d.). Retrieved from http://www.nis.eu/

OAO Rossijskije schelesnyje dorogi. (n.d.). Retrieved from http://rzd.ru/

Radojcic, D. (2012). *New Opportunities on the Danube Corridor*. Transport Research and Innovation Portal. Retrieved from http://www.transport-research.info/sites/default/files/project/documents/20120314_140325_25734_New%20Opportunities%20on%20the%20Danube%20Corridor.pdf

Radovanović, N. (2016). COP 21 and What It Means for a Company in the Oil and Gas Sector – The NIS JSC Perspective. *National Petroleum Committee of Serbia – World Petroleum Council Bulletin, 21*(September), 17–20.

Rajal, B., & Petrović, A. (2016). *Serbia Oil & Gas Regulation 2016*. International Comparative Legal Guides. Retrieved from http://www.iclg.co.uk/practice-areas/oil-and-gas-regulation/oil-and-gas-regulation-2016/serbia

Ramani, S. (2016, February 15). Why Serbia Is Strengthening Its Alliance with Russia. *The World Post*. Retrieved from http://www.huffingtonpost.com/samuel-ramani/why-russia-is-tightening-_b_9218306.html

Reljić, D. (2016). Russia Gives Serbia the Choice: Satellite or Bargaining Chip. In S. Fischer and M. Klein (eds.), *Conceivable Surprises Eleven Possible Turns in Russia's Foreign Policy* (pp. 25–29). SWP Research Paper 10. Berlin: Stiftung Wissenschaft und Politik. Retrieved from http://www.swp-berlin.org/fileadmin/contents/products/research_papers/2016RP10_fhs_kle.pdf

Ristić, S. (2011). *Oil Derivatives Market in Republic of Serbia*. Presentation of the Ministry of Infrastructure and Energy of the Republic of Serbia, 3rd Energy Community Oil Forum, 27–28 October, Belgrade. Retrieved from https://www.energy-community.org/portal/page/portal/ENC_HOME/DOCS/12 04187/0633975AB5787B9CE053C92FA8C06338.PDF

Savic, M. (2012, March 13). Comico Oil Wins Permit to Build $250 Million Refinery in Serbia. *Bloomberg*. Retrieved from http://www.bloomberg.com/news/articles/2012-03-13/comico-oil-wins-permit-to-build-250-million-refinery-in-serbia

Sekularac, I. (2014, August 12). UPDATE 1-Serbia to Investigate Privatisation of State Oil Firm NIS. *Reuters*. Retrieved from http://uk.reuters.com/article/serbia-nis-idUKL6N0QI3BZ20140812

Serbia, Russia Initial Deal on Oil Monopoly's Sale. (2008, December 22). *Reuters*. Retrieved from http://uk.reuters.com/article/energy-serbia-russia-idUKLM10432220081222

Serbia Suspects Russian Oil Giant Lukoil of Trade Violations. (2013, October 17). *Russian Legal Information Agency*. Retrieved from http://rapsinews.com/news/20131017/269262249.html

Serbia's Smederevo City Agrees to Settle Spat with Comico Oil Out of Court – Media. (2012, December 20). *SeeNews*. Retrieved from https://seenews.com/news/serbias-smederevo-city-agrees-to-settle-spat-with-comico-oil-out-of-court-media-324297

Serbian Oil and Gas Privatization: Investigation Promised. (2014, August 19). *Radio Slobodna Evropa*. Retrieved from http://www.slobodnaevropa.org/a/serbia-oil-and-gas-privatization-investigation-promised/26539837.html

Serbian Town Scraps Deal with U.S.-Dutch Oil Company. (2013, February 5). *B92*. Retrieved July 27, 2016 from http://www.b92.net/eng/news/business.php?yyyy=2013&mm=02&dd=05&nav_id=84529

Souček, I., & Vasiliev, D. (2014). *Investment in Refining Capacities in Serbia*. Presentation of NIS a.d., 6th Energy Community Oil Forum, 30 September, Belgrade. Retrieved from https://www.energy-community.org/portal/page/portal/ENC_HOME/DOCS/3376153/0633975ADCA57B9CE053C92FA8C06338.PDF

Statistical Office of the Republic of Serbia. (2016a). *Energy Balances, 2014*. Belgrade: Statistical Office of the Republic of Serbia. Retrieved from http://webrzs.stat.gov.rs/WebSite/repository/documents/00/02/03/47/SB-607-BILTEN_ENERGETIKE_2014.pdf

Statistical Office of the Republic of Serbia. (2016b). *Statistical Pocketbook of Serbia*. Belgrade: Statistical Office of the Republic of Serbia. Retrieved from http://webrzs.stat.gov.rs/WebSite/repository/documents/00/02/07/28/Statisticki_kalendar_2016.pdf

Statistical Office of the Republic of Serbia. (n.d.). Retrieved from http://webrzs.stat.gov.rs/

Trejbal, V. (2012, October 10). Doprava na Dunaji skomírá. Spojením s Rýnem by vznikla evropská tepna. *Patria Online*. Retrieved from https://www.patria.cz/zpravodajstvi/2168171/doprava-na-dunaji-skomira-spojenim-s-rynem-by-vznikla-evropska-tepna.html

Trifkovic, D. (2016, February 29). The Vucic Government's Secret Pact with NATO. *Fort Russ*. Retrieved from http://www.fort-russ.com/2016/02/the-vucic-governments-secret-pact-with.html

Who Is (Comico Oil) Going to Build Oil Refinery in Smederevo? (2013, January 18). *InvestInSerbia*. Retrieved from http://www.invest-in-serbia.com/general/84938-who-is-going-to-build-oil-refinery-in-smederevo.html

Zimonjic, V. P. (2012, January 19). BALKANS: The Dark Side of Serbia's Oil Shale Fairy Tale. *INTER PRESS SERVICE News Agency*. Retrieved from http://www.ipsnews.net/2012/01/balkans-the-dark-side-of-serbias-oil-shale-fairy-tale/

List of Interviews

Interview 01: Belgrade, Serbia, September 27, 2016.
Interview 02: Belgrade, Serbia, September 28, 2016.
Interview 03: Belgrade, Serbia, September 28, 2016.
Interview 04: Belgrade, Serbia, September 28, 2016.
Interview 05: Belgrade, Serbia, September 29, 2016.
Interview 06: Belgrade, Serbia, September 29, 2016.

CHAPTER 12

Countries with Limited Russian Activities: Kosovo, Montenegro, and Slovenia

12.1 Kosovo

12.1.1 Crude Oil Sector General Information

12.1.1.1 Introduction and Upstream

The Republic of Kosovo is a small landlocked country in the Central Balkans with approximately 1.8 million inhabitants neighbouring Albania, Montenegro, Serbia, and the Republic of Macedonia. It is a partially recognized state that emerged after separation from Serbia in 2008. Serbia still claims the region as the Autonomous Province of Kosovo and Metohija (Kosmet) within the Republic of Serbia. Kosovo does not import crude oil as there are no facilities capable of processing it into petroleum products. Nor are there oil deposits in the Republic of Kosovo, and no oil extraction takes place. All Kosovan consumption must therefore come from petroleum products. The import level is stable at 550,000–600,000 tonnes annually, with diesel accounting for roughly one-half that amount (Energy Regulatory Office, 2011, 2013). Petroleum products represented 26.7% of TPES in 2014 (24.2% in 2013; Kosovo Agency of Statistics, 2015, p. 4).

12.1.1.2 Midstream

There is no pipeline in the Republic of Kosovo to transport either crude oil or petroleum products. Imports come overland by either highway transport (75%; Albania, Bosnia and Herzegovina, Greece, Montenegro,

© The Author(s) 2019
T. Vlček, M. Jirušek, *Russian Oil Enterprises in Europe*,
https://doi.org/10.1007/978-3-030-19839-8_12

Italy, Croatia, and others) or rail (25%; Serbia, Macedonia) (Energy Regulatory Office, 2011, p. 33, 2015, p. 28). Rail is not used to capacity, although the railway system is connected to the common petroleum terminals with loading and unloading facilities (UNMIK-EU Pillar PISG Energy Office, 2005). Petroleum product imports originate in Albania (46%), Bosnia and Herzegovina (35%), Serbia (13%), Greece (3%), and Montenegro (3%) (Energy Regulatory Office, 2015, p. 27). Given the situation in Albania,[1] import levels from the country are expected to drop in favour of imports from other countries.

12.1.1.3   Downstream

There are no refineries in the Republic of Kosovo. The only petroleum product produced domestically is heavy fuel oil for heating in four licenced production plants. This production covers 25% of the demand for heavy fuel oil (Energy Regulatory Office, 2015, p. 26). It is produced from imported oil semiproducts.

Official documents state that the market is highly competitive and that is why no price regulation is necessary (Energy Regulatory Office, 2015, p. 29). And there are remarkably many companies operating on the Kosovan market, especially given its size. The biggest company on the market is Al-Petrol shpk, which operates 49 petrol stations and is the largest wholesaler in Kosovo (*Al-Petrol shpk*, n.d.). Other large companies include Kosova Petrol shpk (35 petrol stations, positioning itself as the leading Kosovan energy company), Ex Fis shpk, HiB Petrol shpk (11 petrol stations), Petrol Group (11 petrol stations), Petrol Company shpk, IP Italian Petrol Kosova shpk, Mamidoil Kosovo, and many others. Other information, however, suggests that the introduction of price regulation would be helpful. The president of the Association of Kosovo Oilmen, Fadil Berjani, recently stated that the market is dominated by four major importers who dictate prices (Shala, 2016). Russian companies have a presence in Kosovo via the retail firm Beopetrol-pristina D.o.o., a subsidiary of the Serbian company Lukoil Srbija a.d. (subsidiary of Russian PAO Lukoil), which controls around 20% of the market.

The Republic of Kosovo's petroleum product market is generally non-transparent. In addition to product quality issues, it is also impacted by illegal operations of a grave nature, including customs fraud (Ex Fis

---

[1] See the chapter on Albania.

shpk, Petrol Company shpk, Al-Petrol shpk, HiB Petrol shpk, etc.), the smuggling of oil from Serbia (Petrol Company shpk, HiB Petrol shpk, Al-Petrol shpk, NTP Gold Benz, etc.), and unpaid rents for point-of-sale fuel outlets (Kosova Petrol shpk). Bedri Selmani, the CEO of Kosova Petrol shpk, managed to sign a contract during the administration of UNMIK to rent the petrol station networks of INA d.d. (Industrija nafte d.d.) and Jugopetrol AD of the former Yugoslavia. Allegedly, he did not pay rent to the Privatization Agency of Kosovo and is currently on trial for the illegal use of 23 Croatian petrol stations owned by INA d.d. He is said to have left Kosovo and now resides in Croatia, where he continues to do business (*'Dogana zbulon'*, 2012; *'"HIB Petrol", "AL Petrol"'*, 2014; *'HIB Petrol prinë'*, 2014; Olluri, 2012; *'Reagon Kosova'* 2016; Veliu, 2016).

12.1.2   Recent Market Developments and Russian Activities

The Russian Federation has no presence in Kosovo's petroleum products market and is not a shareholder in any company which is (Beopetrol-pristina D.o.o. is controlled by PAO Lukoil through the Serbian company Lukoil Srbija a.d.). The country's relations with Russia are very poor—Russia was strongly opposed to Kosovo's declaration of independence from Serbia in 2008, a position attributable to Russia's excellent relations with Serbia and its interest in maintaining the latter's territorial integrity. This reasoning still survives: the Russian Federation does not recognize the Republic of Kosovo. Some argue that Kosovo's declaration of independence from Serbia was recently seized upon by Russia as a pretext for Crimea's declaration of independence from Ukraine, and that Crimea is Russia's vengeance for Kosovo (Barlovac, 2014; Domi, 2014; Gessen, 2014; Radušević, 2015).

12.1.3   Research Indicator Assessment

Russian Federation companies are not present in any way on the Kosovan oil market. Because of this, the research indicator assessment is irrelevant to the case study of Kosovo.

12.2   Montenegro

12.2.1   Crude Oil Sector General Information

12.2.1.1   Introduction and Upstream

Montenegro is a small country in the Western Balkans with approximately 680,000 inhabitants living on less than 18,000 km². It borders Bosnia and Herzegovina, Serbia, Kosovo, and Albania and on the west, the Adriatic Sea. For a country of its size, petroleum product consumption is low—in fact, since it has no refineries, Montenegro imports no crude oil at all. Consumption is thus entirely dependent on petroleum products imported from abroad, mainly from Greece. In 2013, the country consumed 272,000 tonnes of these products, the vast majority in the transport sector (Government of Montenegro Negotiating Team on Accession to the European Union, 2015, p. 8). Petroleum products account for one-third of Montenegro's TPES (Table 12.1).

Oil production in Montenegro is non-existent, but since 2010 the Ministry of Economy has been preparing a tender for crude oil and natural gas exploration off the country's southern Adriatic coast. The ministry states that Montenegro would be capable of covering its oil and gas needs from its own resources (Hastorun, 2015, p. 32.3). Fifteen companies expressed interest in the tender in 2011. But although the tender was closed in 2014, the Ministry of Economy did not start talks with the winning bidder until late 2015 (*'Montenegro Attracts'*, 2011). The winner was an Italian-Russian consortium consisting of Eni S.p.A. and NOVATEK OAO.[2] Montenegro's Minister of Economy, Vladimir

Table 12.1   Montenegro crude oil data

	2010	2011	2012	2013
Consumption	0	0	0	0
Production	0	0	0	0
Import dependency	0%	0%	0%	0%
Consumption of oil derivatives	314	301	282	272

Source: Government of Montenegro Negotiating Team on Accession to the European Union, 2015, p. 8

Note: Figures in thousands of tonnes; percentage calculations by T. Vlček

[2] 9.99% of NOVATEK OAO was owned by OAO Gazprom as of 2010 (Mazneva, 2010). The current share is unknown.

Kavarić, announced that under the contract, the companies will drill three mandatory boreholes and one servicing borehole (Koseva, 2016). The drilling concession, approved by the Parliament in June 2016, was granted for 30 years for a total area of approximately 1200 square km ('*Eni, Novatek to*', 2016).

12.2.1.2   Midstream
There are no pipelines in Montenegro and no crude oil is imported.

12.2.1.3   Downstream
There are no refineries in Montenegro—the country therefore imports 100% of its needs in the form of petroleum products. The dominant company on the market is Jugopetrol AD Kotor, which operates a network of 38 petrol stations under the EKO brand. This gives the company control of 43% of the petrol stations in Montenegro since there are only 89 such outlets in the entire country (Energy Commission, 2013, p. 5). The company also supplies fuel for yachts and airports and controls the largest storage facilities and operates three truck loading facilities and a fleet of vehicles, giving the company control of 60% of the total oil products market (Government of Montenegro Negotiating Team on Accession to the European Union, 2015, p. 6); 54.4% of the company is owned by Hellenic Petroleum International SA, a subsidiary of Greece's Hellenic Petroleum SA (Table 12.2).

Table 12.2   Jugopetrol AD Kotor ownership structure as of 2011

Shareholder	Share (%)
Hellenic Petroleum International SA	54.4
Privatizacioni fond 'MIG'	9.2
Fond zajedničkog ulaganja 'TREND'	6.0
HB Zbirni kastodi racun 3	4.4
Fond zajedničkog ulaganja 'ATLAS MONT'	1.9
CK Zbirni kastodi racun 5	1.5
NM-Zbirni kastodi racun 6	1.2
Other legal entities	6.9
Individuals	14.7
Republic of Montenegro[a]	0.0

Source: Jugopetrol AD Kotor, 2012, p. 6

[a]Republic of Montenegro owns 1 share worth EUR 15

Montenegro Bonus d.o.o. Cetinje is the second biggest wholesaler and also the operator of a gas transmission system. LUKOIL Montenegro d.o.o., a subsidiary of PAO Lukoil, operates a network of 11 petrol stations (12% of the market). Petrol Crna Gora MNE d.o.o. (a subsidiary of Slovenian Petrol d.d.) operates a network of ten petrol stations (11% of the market). The remaining 30 petrol stations are operated by a number of small companies, often with only a single petrol station.

12.2.2   Recent Market Developments and Russian Activities

The relationship between Russia and Montenegro is intensive; but it is based largely around tourism (25% of tourists are Russians), housing (7000 Russians are registered as permanent residents), and industry (32% of Montenegro companies are under Russian ownership) (Kekić, 2015, p. 8; Vagner, 2013). Energy, especially oil and/or petroleum products, is not a major factor. Economic links between Russia and Montenegro tend to be overstated, too. In 2013, only 0.4% of Montenegro's imports came from Russia (compared to 55.3% from the EU) and as little as 0.1% of Montenegro's exports were aimed at Russia (compared to 3.8% to the EU) (Kekić, 2015, p. 9).

Generally speaking, Russia's chief interest in Montenegro is to block its integration into the Euroatlantic structures NATO and the EU (as with other Balkan countries) (Szpala, 2014, p. 1). In December 2015, for example, after the formal invitation to join NATO was issued, Russia reacted with fierce criticism, with Sergey Lavrov invoking the NATO aerial bombing of Yugoslavia in 1999 and calling for a referendum in Montenegro on membership (Ledger, 2016). Yet Montenegro is aware of the exaggeration of its relations with Russia by the media and continues its pro-Western policy (Montenegro joined the EU sanctions regime against Russia in March 2014; voted against Russia in the UN General Assembly resolution on Crimea; Milo Đukanović, the Prime Minister of Montenegro, turned down an invitation to attend World War II Victory Day celebrations in Moscow in 2015, etc.; Samorukov, 2015). Despite its fierce verbal criticism, the Russian reaction did not impact on day-to-day life (in the form of a visa regime, negative media campaign, free trade agreement withdrawal, a variety of bans, etc.). In July 2018, though, it came to light in a report by the Foreign Policy Research Institute that 'Russia coordinated with the local opposition and Serb ethno-nationalists in an unsuccessful attempt to topple the democratically elected government of Montenegro in October 2016' (Bajrović, Garčević, & Kraemer,

2018, p. 5). This confirms the above-mentioned primary foreign policy strategy directed at Montenegro of blocking its integration into NATO and the EU. So far, these efforts have failed—Montenegro joined NATO on 5 June 2017.

As for Russian oil companies on the Montenegro market, the only active presence is that of LUKOIL Montenegro d.o.o., formed in 2008 when PAO Lukoil bought a network of six petrol stations from the domestic firm Rokšped d.o.o. (*'Russians Buy'*, 2008). Even though Lukoil raised that number to 11 stations (12% of the retail market), its position is far from dominant and there are no indications of any strategy to achieve dominance. The evidence instead suggests that the purchase was undertaken in Montenegro for the same reason Lukoil Srbija a.d. bought Beopetrol a.d. and its 113-station network in Serbia (7.8% of the market)—to support PAO Lukoil's position in the Balkan Peninsula by selling products from PAO Lukoil-owned refineries in Bulgaria. The oil sector in the Balkan Peninsula is highly competitive, with a strong presence of companies native to the region, from Greece, Slovenia, and Croatia. In any event, Russia's efforts could hardly be construed as politically motivated. Rather, the evidence shows a purely economic logic at work. Vladimir Repin, former General Director of LUKOIL-Beopetrol A.D., stated in a 2006 interview that 'Montenegro is becoming a major European tourist centre, attracting growing numbers of tourists, including those travelling by car. The standard of living in the region is rising. All this is indicative of the promising nature of the fuel market in Serbia and Montenegro' (*'According to European'*, 2006).

12.2.3   Research Indicator Assessment

Because of the information presented above, the research indicator assessment is not applicable to this case study.

12.3   Slovenia

12.3.1   Crude Oil Sector General Information

12.3.1.1   Introduction and Upstream

The Republic of Slovenia was the northernmost country to emerge from the former Yugoslavia. It neighbours Italy, Austria, Hungary, and Croatia and has access to the Adriatic Sea limited to just 10 km of coastline

Table 12.3   Republic of Slovenia crude oil data

	2008	2009	2012	2014
Consumption	2846	2475	2454	2224
Production	0.158	0.134	15.2	–
Import dependency	100%	100%	99.4%	–

Source: Index Mundi, n.d.; Republic of Slovenia Statistical Office, n.d.

Note: Total final consumption of all petroleum products. Figures in thousands of tonnes; percentage calculations by T. Vlček

connecting the cities of Portorož, Izola, and Koper. Two million inhabitants live in an area of around 20,000 km². The country has been an EU member since 2004.

Since 1998, when oil processing operations were terminated at the refinery in Lendava, the country has been fully dependent on petroleum product imports. Until that year, production from the Lendava refinery covered approximately 20% of Slovenia's needs. Slovenian consumption of petroleum products is around 2.2–2.4 mta annually, making up approximately 35% of gross inland energy consumption (Table 12.3).

Oil extraction is conducted exclusively by Geoenergo d.o.o., a company owned by Nafta Lendava, d.o.o. (50%) and Petrol d.d. (50%). Oil is exploited from the Petišovici field in the Mura Depression and is of negligible volume. Geoenergo d.o.o. has entered into a Joint Operating Agreement with Ascent Slovenia Limited, a subsidiary of British Ascent Resources plc. All operational activity is actually undertaken by Ascent Resources d.o.o. (Geoenergo d.o.o., n.d.); Petrol d.d., the co-owner of Geoenergo d.o.o., is the Slovenian oil champion, and, in addition to Slovenia, it operates in Cyprus (Cypet Oils, Ltd.), Austria (Petrol Trade GmbH), Croatia (Petrol Hrvatska d.o.o.), Bosnia and Herzegovina (Petrol Trgovina, d.o.o. and Petrol BH Oil Company, d.o.o.), Montenegro (Petrol Crna Gora MNE d.o.o.), and Serbia (Table 12.4).

Petrol d.d. states that the petroleum products sold in Slovenia are imported on the basis of procurement procedures but at the same time the country has increased the share of procurement from local inland refineries located in SE Europe (Petrol d.d., 2016, p. 51). Therefore, only small petroleum product amounts are imported from Russia: 199,000 tonnes in 2013; 256,000 tonnes in 2014, and 93,000 tonnes in 2015 (Republic of Slovenia Statistical Office, n.d.). In 2015, Petrol d.d. sold 3 million tonnes

Table 12.4 Ownership structure of Petrol d.d.

Shareholder	Share (%)
Slovenski Državni Holding, d.d.	19.75
Československá obchodní banka a.s. (FID)	12.79
Kapitalska Družba, d.d.	8.27
Vizija Holding, k.d.d.	3.44
Vizija Holding Ena, k.d.d.	3.05
NLB d.d.	3.03
Nova KBM d.d.	2.06

Source: Petrol d.d., 2016, p. 21

Note: The table lists shareholders with a minimum of 2% of shares in the company

of petroleum products (Petrol d.d., 2016, p. 48); Russian products thus made up only 3.1% of Petrol d.d.'s sales. The statistics are even less impressive for petroleum product imports from Serbia, Bosnia and Herzegovina, Bulgaria, and Romania (where the refineries are owned by Russian entities).

12.3.1.2 Midstream

Crude oil was imported to Slovenia's refinery in Lendava via the JANAF system[3] through a small branch diverting off the main stretch at Virje. Petroleum products are imported via the Adriatic Sea through the port of Koper (90%) and by rail (10%).

12.3.1.3 Downstream

There is only one refinery in Slovenia, located in Lendava. That refinery, owned by Eko-Nafta, proizvodnja naftnih derivatov d.o.o. (a 100% subsidiary of Nafta Lendava, d.o.o.), was closed in July 1998 due to poor economic performance and has been for sale ever since. Nafta Lendava, d.o.o. also operated a methanol plant with a capacity of 165,000 tonnes annually. This operation went bankrupt in 2014 and the next year, in 2015, was purchased at auction by US Methanol Corporation for EUR 5.6 million ('*US Buyer*', 2016; '*US Firm*', 2015) (Table 12.5).

Nafta Lendava, d.o.o. is owned by the Republic of Slovenia via Slovenian Sovereign Holding. It has been insolvent since 2014 and generates no profit, but it does have stakes in other companies, including 100% of Nafta

---

[3] See detailed information about the JANAF system in the chapter on Croatia.

Table 12.5   Capacity of Slovenian refinery as of 2016

Refinery	Owner	Refining capacity	Nelson complexity index (2016)	Year established
Lendava	Eko-Nafta, proizvodnja naftnih derivatov d.o.o.	0.672 mta	–	1964

Source: Compiled by T. Vlček from public sources

varovanje in požarna varnost, d.o.o.; 100% of Eko-Nafta, proizvodnja naftnih derivatov d.o.o.; and 50% of Geoenergo, d.o.o. (Slovenian Sovereign Holding, 2016, p. 75).

There are around 540 petrol stations in Slovenia, of which Petrol d.d. operates 316, giving it a 58% share of the retail market (Petrol d.d., 2016, p. 50). OMV Slovenija, d.o.o. operates a network of 106 petrol stations (20% of the market). Petrol Lukoil d.o.o., a subsidiary of PAO Lukoil, operates 52 petrol stations. MOL Slovenija, d.o.o. has 40 stations (7% of the market). MOL Slovenija, d.o.o. purchased in July 2016 the 17 petrol station network of Eni Slovenija, družba za trženje z Naftnimi Derivati, d.o.o. thereby increasing the number of its petrol stations to 40. Slovenian INA—Industrija nafte d.d. operates 6 petrol stations, while a number of smaller companies operate the remaining circa 70 stations in the Slovenian market.

12.3.2   Recent Market Developments and Russian Activities

PAO Lukoil was interested in purchasing the refinery in Lendava, the port operator Luka Koper d.d. (with plans to build an oil logistics centre there for its business operations in the region), and Petrol d.d. Vagit Alekperov, CEO of PAO Lukoil, met Andrej Vizjak (Slovenia's Minister of Economic Development and Technology) and Janez Janša (Slovenia's Prime Minister) to discuss the idea in 2006 ('*LUKoil in Talks*', 2006). Negotiations, however, took a different direction. Later in 2006, PAO Lukoil and Petrol d.d. signed a framework agreement to create a joint venture for oil product sales. This venture, in which PAO Lukoil was to have the 49% minority share, anticipated merging Petrol d.d.'s stations in Slovenia with the networks of its subsidiaries in Croatia and Bosnia and Herzegovina. PAO Lukoil was expected to add Lukoil Macedonia DOOEL Skopje and Beopetrol a.d. (later Lukoil Srbija a.d.) ('*LUKoil and*

Slovenian', 2006; '*Petrol and Lukoil'*, 2006). This unification of the petrol stations of these two companies throughout the former Yugoslavia would benefit both firms by expanding their retail markets. PAO Lukoil was especially interested in selling products from its refineries in Bulgaria and Romania in Petrol d.d.'s subsidiaries spread across the Western Balkans: besides the over 300 petrol stations in Slovenia, it also targeted the networks of Petrol Hrvatska d.o.o. (102 petrol stations in Croatia), Petrol BH Oil Company, d.o.o. (35 petrol stations in Bosnia and Herzegovina), and Petrol Crna Gora MNE d.o.o. (10 petrol stations in Montenegro).[4] Petrol d.d.'s main interest lay in long-term stable access to crude oil through PAO Lukoil. But because the investment positions of the two companies were at variance, the plan was scrapped just one year later (*'Slovenia's Petrol'*, 2007). The main target of PAO Lukoil was allegedly access to the Croatian market, but shortly after the deal with Petrol d.d. fell through, PAO Lukoil entered the Croatian market on its own. In 2008, Lukoil Europe Holdings (a 100% subsidiary of PAO Lukoil) purchased the 9-petrol station network of Europa Mil d.o.o., which it then used to develop its position into a 52-station network by 2016.

A similar story may be told regarding PJSC Gazprom Neft. In 2011, Alexander Dyukov, Chairman of the Management Board of PJSC Gazprom Neft, and Tomaž Berločnik, Chairman of the Management Board of Petrol d.d., signed a Memorandum of Understanding during Vladimir Putin's official visit to Slovenia (*'Gazprom Neft and Slovenian'*, 2011, n.d.). The agreement was aimed at cooperative ventures in the Balkan oil market, especially combining PJSC Gazprom Neft's refining capacities and markets with Petrol d.d.'s markets. PJSC Gazprom Neft's business interests in Slovenia and Croatia are visible in the agreement, but nothing concrete has come of the Memorandum so far.

12.3.3   Research Indicator Assessment

Two indicators were confirmed in Slovenia, namely 'active support by Russian state representatives for energy enterprises and their activities abroad' and 'efforts to take control of the energy resources, transit routes, and distribution networks of the client state'. Officials in the Kremlin have supported Russian oil interests in Slovenia; however, the actions of PJSC Gazprom Neft and PAO Lukoil have never specifically targeted Slovenia.

---

[4]Numbers of petrol stations as of 2016; 2006 numbers may differ.

The goal, and it is clearly commercial in nature, has concerned Petrol d.d.'s valuable networks abroad, especially in Croatia but also in Bosnia and Herzegovina and Montenegro. It is the Croatian market that is primarily attractive to Russian oil companies.[5] Even though PAO Lukoil was interested in purchasing the refinery in Lendava (with Petrol d.d.'s petrol station network), commercial calculations prevailed, and the idea was abandoned because of the abysmal condition of the Lendava refinery. Nevertheless, all activities in the Slovenian oil sector have been carried out with economic interests foremost—centred on expanding retail markets in the Balkan Peninsula to allow for the sale of the products from Russian-owned refineries in the region (Table 12.6).

Table 12.6   Summary of indicators for Slovenia

Indicator	Found	Found with
Active support by Russian state representatives for energy enterprises and their activities abroad	Inconclusive	PJSC Gazprom Neft, PAO Lukoil
As a foreign supplier, Russia rewards certain behaviours and links energy deals to the client state's foreign policy orientation	No	–
Abuse of infrastructure (e.g. pipelines) and differential pricing to exert pressure on the client state	No	–
Efforts to take control of the energy resources, transit routes, and distribution networks of the client state	Inconclusive	PAO Lukoil
Disruption (by various means) of alternative supply routes/sources of supply	No	–
Efforts to gain a dominant market position in the client country	No	–
Efforts to eliminate competitive suppliers	No	–
Acting against liberalization	No	–
Diminishing the importance and influence of multilateral regimes such as the EU	No	–
Attempts to control the entire supply chain (regardless of commercial rationale)	No	–
Economically irrational steps taken to maintain a particular position in the client state's market	No	–

Source: Author

Note: *Inconclusive* means some indications of this behaviour were found, but not of a shape, size, or importance to be ascribed to strategic behaviour and thus fulfil the indicator

[5] See the chapter on Croatia for detailed information.

SOURCES

According to European Standards. (2006). *Oil of Russia, 2.* Retrieved from http://www.oilru.com/or/27/477/

Al-Petrol shpk. (n.d.). Retrieved from http://www.alpetrol-ks.com/

Bajrović, R., Garčević, V., & Kraemer, R. (2018). *Hanging By A Thread: Russia's Strategy of Destabilization in Montenegro.* Philadelphia: Foreign Policy Research Institute. Retrieved from https://www.fpri.org/wp-content/uploads/2018/07/kraemer-rfp5.pdf

Barlovac, B. (2014, March 18). Putin Says Kosovo Precedent Justifies Crimea Secession. *BalkanInsight.com.* Retrieved from http://www.balkaninsight.com/en/article/crimea-secession-just-like-kosovo-putin

Dogana zbulon katër kompani që i janë shmangur taksave. (2012, December 14). *koha.net.* Retrieved from http://koha.net/?id=8&arkiva=1&l=127393

Domi, T. L. (2014, March 24). Putin Exacts His Revenge for the "Brothers Across the Danube"; Is Kosovo Crimea Now? *Emerging Democracies Institute.* Retrieved from http://edi-dc.org/putin-exacts-revenge-brothers-across-danube-kosovo-crimea-now/

Energy Commission. (2013). *Screening Report Montenegro: Chapter 15 – Energy.* Retrieved from http://ec.europa.eu/enlargement/pdf/montenegro/screening_reports/screening_report_montenegro_ch_15.pdf

Energy Regulatory Office. (2011). *Statement of Security of Supply for Kosovo (Electricity, Natural Gas and Oil).* Prishtinë: Energy Regulatory Office. Retrieved from https://www.energy-community.org/pls/portal/docs/1218181.PDF

Energy Regulatory Office. (2013). *Statement of Security of Supply for Kosovo (Electricity, Natural Gas and Oil).* Prishtinë: Energy Regulatory Office. Retrieved from https://www.energy-community.org/portal/page/portal/ENC_HOME/DOCS/2422181/0633975AD4417B9CE053C92FA8C06338.PDF

Energy Regulatory Office. (2015). *Statement of Security of Supply for Kosovo (Electricity, Natural Gas and Oil).* Prishtinë: Energy Regulatory Office. Retrieved from https://www.energy-community.org/portal/page/portal/ENC_HOME/DOCS/3842352/2198FDD2CC0E1492E053C92FA8C0BE17.PDF

Eni, Novatek to Search for Oil Offshore Montenegro. (2016, September 14). *Offshore Energy Today.* Retrieved from https://www.offshoreenergytoday.com/eni-novatek-to-search-for-oil-offshore-montenegro/

Gazprom Neft and Slovenian Petrol Sign Memorandum of Understanding. (2011, March 22). *Gazprom Neft News.* Retrieved from http://www.gazprom-neft.com/press-center/news/3912/

Gazprom Neft and Slovenian Petrol Sign Memorandum of Understanding. (n.d.). *NIS Gazprom Neft Press Release.* Retrieved from http://ir.nis.eu/no_cache/news-and-events/single-news/article/27/

Geoenergo d.o.o. (n.d.). Retrieved from http://www.slovenski-plin.si/

Gessen, M. (2014, March 21). Crimea Is Putin's Revenge. *Slate.com.* Retrieved from http://www.slate.com/articles/news_and_politics/foreigners/2014/03/putin_s_crimea_revenge_ever_since_the_u_s_bombed_kosovo_in_1999_putin_has.html

Government of Montenegro Negotiating Team on Accession to the European Union. (2015). *Action Plan for Implementation of Directive Imposing Obligation on Member States to Maintain Minimum Stocks of Crude Oil and/or Oil Products.* Retrieved from http://www.eu.me/

Hastorun, S. (2015). *2013 Minerals Yearbook – Montenegro.* U.S. Geological Survey. Retrieved from http://minerals.usgs.gov/minerals/pubs/country/2013/myb3-2013-mj.pdf

"HIB Petrol", "AL Petrol", "Petrol Company" Oil Smuggling. (2014, December 11). *Friends of Kosovo.* Retrieved from https://friendsofkosovo.wordpress.com/2014/12/11/hib-petrol-al-petrol-petrol-company-oil-smuggling/

HIB Petrol prinë me kontributet në shtet. (2014, December 31). *lajmi.net.* Retrieved from http://lajmi.net/hib-petrol-prine-me-kontributet-ne-shtet/

Index Mundi. (n.d.). Retrieved from http://www.indexmundi.com/

Jugopetrol AD Kotor. (2012). *2011 Annual Business Report.* Kotor: Jugopetrol AD Kotor.

Kekić, L. (2015). *Presentation at the Russia in the Balkans Conference*, March 13, 2015, The Shaw Library, London School of Economics, London. Conference Report. Retrieved from http://www.lse.ac.uk/europeanInstitute/research/LSEE/Events/2014-2015/Russia-in-the-Balkans/merged-document.pdf

Koseva, D. (2016, February 5). Eni, Novatek Win Offshore Concession in Montenegro. *BNE Intellinews.* Retrieved from http://www.intellinews.com/eni-novatek-win-offshore-concession-in-montenegro-90078/

Kosovo Agency of Statistics. (2015). *Annual Energy Balance in the Republic of Kosovo for 2014.* Retrieved from http://ask.rks-gov.net/en/

Ledger, R. (2016, April 20). Montenegro: Another Balkan Nation to Experience Unrest. *Global Risk Insights.* Retrieved from http://globalriskinsights.com/2016/04/montenegro-another-balkan-nation-experience-unrest/

LUKoil and Slovenian Company Petrol Plan to Establish a Joint Venture. (2006, August 28). *LUKoil Oil Company Press Release.* Retrieved from http://www.lukoil.com/press.asp?div_id=1&id=2561

LUKoil in Talks to Buy Slovenian Refinery. (2006, June 1). *RBC News.* Retrieved from http://rusenergy.blogspot.cz/2006/06/lukoil-in-talks-to-buy-slovenian.html

Mazneva, Y. (2010, December 29). Novatek's Largest Shareholder Is Its CEO. *The Moscow Times.* Retrieved from https://themoscowtimes.com/articles/novateks-largest-shareholder-is-its-ceo-4057

Montenegro Attracts 15 Firms for Oil Concessions. (2011, February 21). *Reuters.* Retrieved from http://www.energy-pedia.com/news/montenegro/montenegro-attracts-15-firms-for-oil-concessions

Olluri, P. (2012, February 21). Kosova Petrol "Occupying" Croatian Property. *BalkanInsight.com.* Retrieved from http://www.balkaninsight.com/en/article/kosova-petrol-occupying-croatian-property

Petrol and Lukoil Together on the Oil Markets of Central and SE Europe. (2006, August 28). *Petrol Press Release.* Retrieved from http://www.petrol.eu/sites/default/files/attachment/petrol_lukoil_28avgust06_ang.pdf

Petrol d.d. (2016). *Annual Report 2015.* Ljubljana: Petrol d.d.. Retrieved from http://www.petrol.eu/sites/www.petrol.eu/files/attachment/annual_report_petrol_2015.pdf

Raduševič, M. (2015, October 23). Velmocenská hra o Kosovo. Jelcinova zrada a Putinova msta. *Literární noviny.* Retrieved from http://www.literarky.cz/politika/svet/20845-velmocenska-hra-o-kosovo-jelcinova-zrada-a-putinova-msta

Reagon Kosova Petrol. (2016, February 16). *GazetaExpress.com.* Retrieved from http://www.gazetaexpress.com/lajme/reafon-kosova-p-168583/?archive=1

Republic of Slovenia Statistical Office. (n.d.). Retrieved from http://www.stat.si/statweb

Russians Buy Montenegro Petrol Stations. (2008, April 2). *Balkan Insight.* Retrieved from http://www.balkaninsight.com/en/article/russians-buy-montenegro-petrol-stations

Samorukov, M. (2015, December 9). The Montenegro Gambit: NATO, Russia, and the Balkans. *Carnegie Moscow Center.* Retrieved from http://carnegie.ru/publications/?fa=62232

Shala, F. (2016, February 16). Ferid Berjani: Oil Price Might Have Been Manipulated. *GazetaExpress.com.* Retrieved from http://www.gazetaexpress.com/en/interviews/interviste-fadil-berjani-mund-te-kete-kurdisje-dhe-manipulime-me-cmimin-e-naftes-71698/

Slovenia's Petrol, Russia's Lukoil Scrap Plan for Establishing Joint Company – Petrol Statement. (2007, December 28). *SeeNews.* Retrieved from https://seenews.com/news/slovenias-petrol-russias-lukoil-scrap-plan-for-establishing-joint-company-petrol-statement-222254

Slovenian Sovereign Holding. (2016). *Annual Report – Management of Capital Assets of the Republic of Slovenia and Slovenian Sovereign Holding for 2014.* Ljubljana: The Republic of Slovenia and Slovenian Sovereign Holding. Retrieved from http://www.sdh.si/doc/Pravni_akti/Eng/Letno%20poro%C4%8Dilo%20upravljanja%202014_ANG.pdf

Szpala, M. (2014). *Russia in Serbia – Soft Power and Hard Interests.* Centre for Eastern Studies (OSW) Commentary Number 150, 27.10.2014. Retrieved from http://www.osw.waw.pl/sites/default/files/commentary_150.pdf

United Nations Interim Administration Mission in Kosovo – EU Pillar Provisional Institutions of Self-Government – Energy Office. (2005). Energy Strategy and Policy of Kosovo, White Paper. Liquid Fuels Development Strategy.

US Buyer Pays Up for Petrochem. (2016, February 22). *The Slovenia Times.* Retrieved from http://www.sloveniatimes.com/us-buyer-pays-up-for-petrochem

US Firm Buys Nafta's Methanol Business. (2015, September 30). *The Slovenia Times.* Retrieved from http://www.sloveniatimes.com/us-firm-buys-nafta-s-methanol-business

Vagner, A. (2013, April 8). Российский след в Черногории. *Радио Свобода.* Retrieved from http://www.svoboda.org/a/24950844.html

Veliu, E. (2016, April 18). Ekskluzive: Bedri Selmani largohet nga Kosova. *Portali "Zëri".* Retrieved from http://zeri.info/ekonomia/86094/ekskluzive-bedri-selmani-largohet-nga-kosova/

CHAPTER 13

Findings

13.1 Summary of Findings

No country anywhere in the world strictly adheres to either the strategic approach or the market approach to energy security as no theoretical model appears in its pure form in the real world. Energy security and policy efforts are always a mixture of the two. Thus it is an inclination towards one end of the scale or the other that bears testimony to the policy approach being taken. In this sense, *Russian state-owned energy corporations active in the oil sector in the Balkan Peninsula countries generally conduct business on the basis of a commercial logic* (Table 13.1).

The Balkan Peninsula is a fairly unimportant trajectory in the foreign (energy) policy of the Russian Federation. Russia does not invest large amounts of political capital in the Balkan countries, with the exception of those, such as Serbia, *Republika Srpska* in Bosnia and Herzegovina, Macedonia, and Greece, that are in some way culturally proximate. *Russian companies operating in these areas take advantage of the warmer acceptance of all things Russian because of Slavonic or Pan-Slavic sympathies*

Some of the findings presented here were taken and further developed from a paper entitled 'Challenges and Opportunities of Natural Gas Market Integration in the Danube Region: The South-West and South-East of the Region as Focal Points for Future Development' (Jirušek & Vlček, 2017) published by Masaryk University Press in 2017.

© The Author(s) 2019
T. Vlček, M. Jirušek, *Russian Oil Enterprises in Europe*,
https://doi.org/10.1007/978-3-030-19839-8_13

Table 13.1 Summary of indicators

Country	Albania		Bosnia and Herzegovina		Bulgaria		Croatia		Greece
Indicator	Found	Found with	Found	Found with	Found	Found with	Found	Found with	Found
Active support by Russian state representatives for energy enterprises and their activities abroad	No	–	No	–	Yes	PAO Lukoil, Yukos Petroleum Bulgaria, Pipeline Consortium Burgas-Alexandroupolis Ltd. (OAO AK Transneft, 33.34%; OAO NK Rosneft, 33.33%; PAO Gazprom Neft, 33.33%)	Yes	PAO Lukoil, PJSC Gazprom Neft, PAO NK Rosneft, OAO Zarubezhneft, PAO Transneft	Yes
As a foreign supplier, Russia rewards certain behaviours and links energy deals to the client state's foreign policy orientation	No	–	Inconclusive	–	No	–	No	–	No
Abuse of infrastructure (e.g. pipelines) and differential pricing to exert pressure on the client state	No	–	No	–	Inconclusive	–	No	–	No
Efforts to take control of the energy resources, transit routes and distribution networks of the client state	No	–	No	–	Inconclusive	–	Yes	PJSC Gazprom Neft, PAO NK Rosneft, OAO Zarubezhneft, PAO Transneft	Yes

Note: Inconclusive means some indications of this behaviour were found, but not of a shape, size, or importance to be ascribed to strategic behaviour and thus fulfil the indicator

	Kosovo		Macedonia		Montenegro		Romania		Serbia		Slovenia	
Found with	Found	Found with	Found	Found with	Found	Found with	Found	Found with	Found	Found with	Found	Found with
Pipeline Consortium Burgas-Alexandroupolis Ltd. (OAO AK Transneft, 33.34%; OAO NK Rosneft, 33.33%; PAO Gazprom Neft, 33.33%), PAO Lukoil	No	–	Incon clusive	PAO Lukoil	No	–	No	–	Yes	–	Incon clusive	PJSC Gazprom Neft, PAO Lukoil
–	No	–	No	–	No	–	No	–	Yes	–	No	–
–	No	–	No	–	No	–	No	–	No	–	No	–
Pipeline Consortium Burgas-Alexandroupolis Ltd. (OAO AK Transneft, 33.34%; OAO NK Rosneft, 33.33%; PAO Gazprom Neft, 33.33%), PAO Lukoil	No	–	No	–	No	–	No	–	No	–	Incon clusive	PAO Lukoil

(continued)

Table 13.1   (continued)

Country	Albania		Bosnia and Herzegovina		Bulgaria		Croatia		Greece
Indicator	Found	Found with	Found	Found with	Found	Found with	Found	Found with	Found
Disruption (by various means) of alternative supply routes/ sources of supply	No	–	No	–	Incon clusive	–	No	–	No
Efforts to gain a dominant market position in the client country	No	–	Yes	OAO Zarubezhneft	Yes	Lukoil Bulgaria Ltd.	Yes	PJSC Gazprom Neft, PAO NK Rosneft, OAO Zarubezhneft	Yes
Efforts to eliminate competitive suppliers	No	–	Incon clusive	–	Yes	Lukoil Bulgaria Ltd.	Incon clusive	–	No
Acting against liberalization	No	–	No	–	Yes	Lukoil Bulgaria Ltd.	No	–	No
Diminishing the importance and influence of multilateral regimes such as the EU	No	–	Yes	–	Yes	–	Incon clusive	–	Yes
Attempts to control the entire supply chain (regardless of commercial rationale)	No	–	No	–	No	–	Incon clusive	PJSC Gazprom Neft, PAO NK Rosneft	No
Economically irrational steps taken to maintain a particular position in the client state's market	No	–	No	–	No	–	No	–	No

Found with	Kosovo		Macedonia		Montenegro		Romania		Serbia		Slovenia	
	Found	Found with	Found	Found with	Found	Found with	Found	Found with	Found	Found with	Found	Found with
–	No	–	No	–	No	–	No	–	Inconclusive	–	No	–
PAO Lukoil	No	–	No	–	No	–	No	–	Inconclusive	–	No	–
–	No	–	No	–	No	–	No	–	No	–	No	–
–	No	–	No	–	No	–	No	–	No	–	No	–
–	No	–	No	–	No	–	No	–	Inconclusive	–	No	–
–	No	–	No	–	No	–	No	–	No	–	No	–
–	No	–	No	–	No	–	No	–	No	–	No	–

and the common Orthodox Church. The positive orientation towards the Orthodox Church and Slavonism in fact forms the basis for Russia's soft power in the Balkan region, and this is the card Russia plays to attain its goals in the Slavonic/Orthodox Balkan countries.

But this familial spirit tends to be *exploited by Russian entities to reach their goals and rarely brings benefit to the Slavonic Balkan countries.* When it comes down to real-world politics, pretensions to Pan-Slavism are abandoned and Russia ceases to behave like a Slavonic brother. As for Greece, the Russian Federation has never had plans to spend political capital there, but rather has sought support for issues elsewhere (i.e. sanctions against Russia, developments in Ukraine, etc.). Meanwhile, Serbia's only tangible benefit from its relationship with Russia has been support over the Kosovo issue. Russia's good relations with *Republika Srpska* came into being during the current era of strong political leadership by President Milorad Dodik. But *Republika Srpska* has never seen any concrete manifestation of support, only verbal backing. For example, loans of EUR 270 million and EUR 500–700 million promised by Russia were never actually made; instead, it is alleged that the loans were used by Milorad Dodik just to gain credit with the voters during a successful election campaign. And this is a pattern: Russian support for its Slavonic brethren is typically oral, not concrete.

Understandably, Russian companies cannot successfully employ soft power everywhere. In fact, only a handful of Balkan countries are without anti-Russian sentiment, permitting a warmer business environment and political relations. Even though the Russian Federation claims the territory occupied by the former Soviet Union as its sphere of influence, where it sets the rules and external actors play a subordinate role (Fischer & Klein, 2016, p. 9), one must remember that no Balkan country has ever been a Union Republic and Cold War relations with major regional players like Yugoslavia and Romania were hardly ever cooperative, let alone comradely. *Thus, the good relationship with some Balkan countries based on the Orthodox Church and Slavonism is primarily the product of relatively modern history, that is, the last 20 to 30 years.* The fall of Yugoslavia and consequent difficulty of development in the region opened the door to tighter relations with Russia.

As an example, the good relations between *Republika Srpska* and the Russian Federation are the product of the Slavonic ethnicity of the majority people (Bosnian Serbs), a common Orthodox religion, historical relations with Russia and Serbia and Pan-Slavism, Russian-peacekeeping troops deployed on the UNPROFOR, IFOR, and SFOR missions in Croatia and Bosnia and Herzegovina, the Russian position on postwar reconstruction, and the good relationship Russia has with the president of *Republika Srpska.*

The good relationship between Serbia and Russia commenced in the late 1990s during the period of Western sanctions against Yugoslavia, when the country relied almost exclusively on Russian imports. It then intensified during the 1999 NATO aerial bombing of Yugoslavia, which was strongly condemned by Russia and brought the two countries even closer. The real background of the relationship between Russia and Serbia is clearly visible in the Kosovo issue, which also led to Russia's entry into the Serbian oil market. Russia's use of the veto in the Security Council to block any UN endorsement of a declaration of independence ('*Gazprom's Bid*', 2008; Sekularac, 2014) and its blocking of Kosovo's membership in the UN has been 'repaid' in the framework energy agreement between the two countries signed in January 2008. There were several points agreed in this framework, including that Serbia would, among other things, assist in construction of the South Stream gas pipeline through the country (in addition to securing the land for the pipeline) and would help build an underground gas storage facility in Banatski Dvor. The sale of Naftna industrija Srbije a.d. was also allowed, in which PJSC Gazprom Neft acquired a 51% stake for EUR 400 million in 2008 along with a 51% share of a domestic company charged with the management and construction of the South Stream pipeline within Serbia (Markovic, 2016; '*Serbian Oil and Gas*', 2014). The suspiciously cheap sale price led to the creation in 2014 of a special investigation team under Interior Minister Nebojša Stefanović to examine the circumstances of NIS a.d.'s privatization, but as of August 2018, the process was still in its preliminary investigation phase.

The good relations the Russian Federation enjoys with Greece are based mainly on military cooperation, Russian support in dealing with Turkey and other Greek political interests, and, of course, on Greece's actions in the energy sphere, where it has promoted various Russian interests inside the EU (Hegedüs, 2010, p. 3). Greece is also a subject of Russian interest for its relations with the Balkan countries and its regional geopolitical position. But the extent of this relationship is often overstated and misunderstood. While it is true that Greece is interested in good relations with the Russian Federation given Russia's investment in sports, real estate, tourism, and the economic residence programme, and it is true that the countries are close because Russia serves as a counterweight against

Turkey, backs Greece on the Cyprus issue, has a common religion, and is culturally and historically proximate, one must be careful to distinguish Greece's self-promoted image on these matters from reality. Greece is in fact more pro-Western, as an examination of its practical political positions shows. Its relations with Russia are instead based on diplomatic gestures of good will and the behaviour and statements of individual politicians who wish to secure good relations for Greece with both the EU and Russia. For example, EU sanctions against Russia were not vetoed by Greece and all official EU statements during the Ukraine crisis were officially supported by the Greek government. The Russian Federation's foreign policy towards Greece is aimed not at Greece itself, but rather at securing support on issues elsewhere (i.e. sanctions against Russia, developments in Ukraine, etc.).

The good relations between the Russian Federation and Bulgaria have different roots. They might harken back to the close relationship between Sofia and Moscow during the communist era, or they may be a simple outgrowth of Bulgaria's deep dependence on Russian energy (90% of natural gas, more than 80% of crude oil, and complete dependence for nuclear fuel). But other pertinent factors include Russian tourism, Russian real property ownership, and the bilateral trade connection (more than 68,000 Russian tourists visited the country in 2013, more than 300,000 Russians own real estate in Bulgaria, and Russian companies have invested more than US $2 billion in Bulgaria; Davydenko, 2014). Russian-Bulgarian relations, though, vary considerably depending upon who is in charge of the Bulgarian government. Centre-right governments are far more cautious when it comes to Russian involvement in Bulgaria.

Russian companies are involved in refining operations in the Balkan Peninsula—at the Brod refinery in Bosnia and Herzegovina (OAO Zarubezhneft, 100% of the country's refining capacity), the Burgas refinery in Bulgaria (PAO Lukoil, 100% of the country's refining capacity), the Petrotel refinery in Romania (PAO Lukoil, 20% of the country's refining capacity), and the Novi Sad and Pančevo refineries in Serbia (PJSC Gazprom Neft, 100% of the country's refining capacity). *We cannot, however, state that the purchase of these refineries has always been motivated by political strategy.*

In Bosnia and Herzegovina, for example, the country closed a direct deal with OAO Zarubezhneft in 2007 upon the initiative of the *Republika Srpska* diaspora in the Russian Federation. Even though the state-owned company made an individual business decision, it had to struggle through complicated negotiations with the Kremlin for state credit to conduct the acquisition (Tepavcevic, 2015, pp. 46, 48). In Bulgaria, two Russian companies, PAO Lukoil and Yukos Petroleum Bulgaria, actually competed against each other, bidding up the price of the asset. Although the Russian government provided strong support, the effort was uncoordinated and shifted from initially backing for Yukos Petroleum's Bulgaria-led consortium to later support for PAO Lukoil 1999 (Vatansever, 2006, pp. 21, 30). In Romania, the decision to purchase a refinery came from privately owned PAO Lukoil, which sought refining capacity to develop its business. In 1997–98, PAO Lukoil was chosen to privatize the Petrotel refinery.

A truly strategic decision with broader geopolitical impact involved the case of Serbia. The country had been trying to privatize Naftna Industrija Srbije a.d. even before the entry of PJSC Gazprom Neft in 2008. The Serbian government hired a consortium of Raiffeisen Investment and Merrill Lynch as consultants for the privatization in 1999. The consultants suggested privatizing 49% of the company in the first phase. Several companies expressed interest, including PAO Lukoil, MOL Rt, OMV AG, Hellenic Petroleum S.A., Petrol d.d., and BP plc (formerly British Petroleum plc). But the public procurement procedure was ultimately unsuccessful; no investor came forward. It was only when the South Stream project was introduced in 2007 that the Russian Federation became interested in entering into a framework energy agreement between the two countries. PJSC Gazprom Neft's acquisition of a 51% stake in NIS a.d., along with other benefits, was repaid by Serbia's agreement to assist in constructing the South Stream gas pipeline over its territory (and securing the land for the pipeline) and in building the underground gas storage facility in Banatski Dvor.[1]

As noted above, Pan-Slavic sentiment surfaced once again after the Yugoslav wars, as did the feelings generated by having in common the Orthodox Church. It is clear that the storyline behind OAO Zarubezhneft's and PJSC Gazprom Neft's entry into the Balkan oil game started with the Yugoslav wars and was motivated by the host countries' quest for foreign investment, while PAO Lukoil's actions in Bulgaria and Romania were market-driven and aimed at developing the company's portfolio with refineries. Within such an economic environment, negotiations with OAO Zarubezhneft and PJSC Gazprom Neft understandably brought different benefits, offsets, and concessions. Russia's involvement, especially in Serbia

---

[1] See Country Case Study: Serbia for detailed information.

and Bosnia and Herzegovina, is positively perceived to this day because it helped to revive a dying industrial subsector and to save jobs for employees in rather underdeveloped regions.

When Russian companies get involved in the oil sector of any Balkan country, they do not target that country in particular, but rather *see it in the context of their broader agenda or a larger project*. The Balkan countries thus serve primarily as channels for those larger projects. This is an unfortunate part of the history of the Balkan Peninsula (Weithmann, 1996).

For example, the Russian targets in Croatia (PAO NK Rosneft, OAO Zarubezhneft, PAO Transneft, PJSC Gazprom Neft) were to use the Adria and JANAF pipeline systems to export Russian oil through the Omišalj terminal; PAO NK Rosneft stated its main interest was JANAF's Omišalj terminal for petroleum export to the Adriatic Sea and its infrastructure, which could be used to deliver oil to the USA. Another target was to develop the company's position in Croatia before it entered the European Union and later benefitted from access to the EU market. It is clear that Russian representatives' actions aimed at penetrating the Croatian energy sector greatly intensified when the exact date of Croatia's accession to the EU was made public. Nikolai Brunich visited Zagreb in January 2013; Alexander Dyukov, Aleksandr Medvedev, and Alexei Miller in January and February 2013; Igor Sechin in June 2013. The frequency of different meetings with Russian officials (or Russian state-owned company officials) dropped considerably once Croatia became part of the European Union in July 2013. Angela Stent reasons that these efforts at bilateral relations with individual member states were made to allow an end-run around what would otherwise be complicated EU bureaucratic procedures (Stent, 2007, p. 426). Russian companies made a substantial effort to penetrate the Croatian market before it became more complicated with Croatia's accession to the EU.

Speaking about Bosnia and Herzegovina, according to a former OAO Zarubezhneft official, the main motive for investment in oil capacities was the company's easier access to customers in the EU markets (Tepavcevic, 2015, p. 44).

The Russian engagement in Bulgaria particularly targeted the AMBO project to strengthen exports of Russian crude oil or crude produced within the Caspian Pipeline Consortium. The Caspian Pipeline Consortium (CPC) is a company and oil pipeline that starts in Tengiz, Kazakhstan, and passes through Kazakhstan and Russia to the Russian Black Sea maritime port of Novorossiysk. The Russian Federation is present in the CPC via

OAO AK Transneft (31% in CPC; it manages the 24% stake for the Russian Federation and a 7% stake for the Russian CPC Company); LukArco B.V. (12.5% in CPC, a 100% subsidiary of PAO Lukoil); and Rosneft Shell Caspian Ventures Ltd. joint venture (7.5% in CPC; OAO NK Rosneft holds 51% in the joint venture and Royal Dutch Shell plc 49%) (Caspian Pipeline Consortium, n.d.; Chevron Corporation, 2015, pp. 23–24; Lukoil Overseas, n.d.; OAO NK Rosneft, n.d.). Altogether the Russian share in CPC is 47.3%.

However, the most common goal of Russian investment in the Balkan oil sector is to secure sales of the company's own products. The development of retail by individual Russian companies operating in the region is crucial to their survival.

Russian companies are involved in the retail market in Bosnia and Herzegovina (OAO Zarubezhneft, 24% of the market, and PJSC Gazprom Neft, 10% of the market), Bulgaria (PAO Lukoil, 7% of the market, and PJSC Gazprom Neft, 1% of the market), Croatia (PAO Lukoil, 6% of the market), Kosovo (PAO Lukoil, 20% of the market), Macedonia (PAO Lukoil, 11% of the market), Montenegro (PAO Lukoil, 12% of the market), Romania (PAO Lukoil, 15% of the retail market, and PJSC Gazprom Neft, 0.9% of the market), and Serbia (PJSC Gazprom Neft, 24% of the market; and PAO Lukoil, 8% of the market) (Table 13.2).

The Balkan region is especially competitive, with several strong local national champions (such as Croatia-based INA—Industrija nafte d.d., Greece-based Hellenic Petroleum S.A., Romania-based The Rompetrol Group N.V., Hungary-based MOL Rt, or Slovenia-based Petrol d.d.) that compete with Russian companies AO NeftegazInKor, Naftna industrija Srbije a.d., and different subsidiaries of PAO Lukoil and PJSC Gazprom Neft. Given the fact that the price of final products rises with increasing distance from the refinery, it is vital for every oil company in the Balkan region to purchase or develop adequate retail chains. Control of both a refinery and a retail network within a reasonable distance guarantees being able to sell your product at the highest possible margin. As the numbers above demonstrate, taking 'control' of a country's market is not on the table; indeed, it is likely impossible given the tough competition from national champions. The frequent purchasing of retail networks and the creation of new ones on the Balkan market are reflective of this strenuous competition. *The actions of Russian companies within the retail market are strongly motivated by an economic logic.*

Table 13.2 Russian involvement in the downstream networks of the SEE countries as of 2016

Country	Company	Russian owner	Number of petrol stations	Share of the market
Albania	–	–	–	–
Bosnia and Herzegovina	Nestro Petrol a.d. Banja Luka	OAO Zarubezhneft	86	24%
	G Petrol d.o.o. Sarajevo	OAO Gazprom	36	9.9%
Bulgaria	Petrol Holding AD	Kirsan Ilyumzhinov[a]	500	15.7%
	Lukoil Bulgaria Ltd.	PAO Lukoil	220	6.9%
	NIS Petrol EOOD	OAO Gazprom	35	1.1%
Croatia	LUKOIL Croatia d.o.o.	PAO Lukoil	52	6.4%
Greece	–	–	–	–
Kosovo	Beopetrol-pristina D.o.o.	PAO Lukoil	?	20%
Macedonia	Lukoil Macedonia DOOEL Skopje	PAO Lukoil	28	11%
Montenegro	LUKOIL Montenegro d.o.o.	PAO Lukoil	11	12%
Romania	S.C. Lukoil Romania S.R.L.	PAO Lukoil	307	15%
	NIS Petrol s.r.l.	OAO Gazprom	18	0.9%
Serbia	NIS Petrol EOOD	OAO Gazprom	346	24%
Slovenia	Petrol Lukoil d.o.o.	PAO Lukoil	52	9.6%

[a]Kirsan Ilyumzhinov is the former president of the Republic of Kalmykia, a Southwestern federal entity of the Russian Federation, and he owns 52.5% of the shares through his Switzerland-based company Credit Mediterranee. The remaining 47.5% is owned by Bulgarian businessman Mitko Sabev

Sources

Caspian Pipeline Consortium. (n.d.). Retrieved from http://www.cpc.ru/

Chevron Corporation. (2015). *Supplement to the Annual Report.* San Ramon: Chevron Corporation. Retrieved from http://www.chevron.com/documents/pdf/annual-report-supplement-2015.pdf

Davydenko, A. (2014, August 8). *Bulgaria-Russia: The Past, the Present, and the Future.* On the 135th Anniversary of the Establishment of Diplomatic Relations Between the Countries. *International Affairs.* Retrieved from http://en.inter-affairs.ru/experts/542-bulgaria-russia-the-past-the-present-and-the-future-on-the-135th-anniversary-of-the-establishment-of-diplomatic-relations-between-the-countries.html

Fischer, S., & Klein, M. (2016). Introduction: Conceivable Surprises in Russian Foreign Policy. In S. Fischer & M. Klein (Eds.), *Conceivable Surprises Eleven Possible Turns in Russia's Foreign Policy* (pp. 5–10). SWP Research Paper 10. Berlin: Stiftung Wissenschaft und Politik. Retrieved from http://www.swp-berlin.org/fileadmin/contents/products/research_papers/2016RP10_fhs_kle.pdf

Gazprom's Bid for Serbia's NIS Brings Pros and Cons for Europe. (2008, January 18). *IHS Markit*. Retrieved from https://www.ihs.com/country-industry-forecasting.html?ID=106597224

Hegedüs, K. (2010). Russia's Relations: The Turkish-Greek-Cypriot Triangle. *International Relations Quarterly 1*(2, Summer), Retrieved from http://www.southeast-europe.org/pdf/02/DKE_02_A_W_Hegedus-Krisztina.pdf

Jirušek, M., & Vlček, T. (2017). *Challenges and Opportunities of Natural Gas Market Integration in the Danube Region. The South-west and South-east of the Region as Focal Points for Future Development*. Brno: Masaryk University. ISBN 978-80-210-8750-7. Retrieved from https://munispace.muni.cz/library/catalog/book/945

Lukoil Overseas. (n.d.). Retrieved from http://lukoil-overseas.com/

Markovic, V. (2016, April 25). Privatization of Serbian Oil Company NIS by Gazprom Still in Pre-Investigation Phase. *Mining See*. Retrieved from http://miningsee.eu/privatization-of-serbian-oil-company-nis-by-gazprom-still-in-pre-investigation-phase/

OAO NK Rosneft. (n.d.). Retrieved from http://www.rosneft.com/

Sekularac, I. (2014, August 12). UPDATE 1-Serbia to Investigate Privatisation of State Oil Firm NIS. *Reuters*. Retrieved from http://uk.reuters.com/article/serbia-nis-idUKL6N0QI3BZ20140812

Serbian Oil and Gas Privatization: Investigation Promised. (2014, August 19). *Radio Slobodna Evropa*. http://www.slobodnaevropa.org/a/serbia-oil-and-gas-privatization-investigation-promised/26539837.html

Stent, A. (2007). *Reluctant Europeans: Three Centuries of Russian Ambivalence Toward the West*. In R. Legvold (Ed.), *Russian Foreign Policy in the 21st Century and the Shadow of the Past* (pp. 393–492). New York: Columbia University Press.

Tepavcevic, S. (2015). The Motives of Russian State-owned Companies for Outward Foreign Direct Investment and Its Impact on State-company Cooperation: Observations Concerning the Energy Sector. *Transnational Corporations, 23*(1), 29–58. Retrieved from http://unctad.org/en/PublicationChapters/diaeia2015d1a2_en.pdf

Vatansever, A. (2006). *Russian Involvement in Eastern Europe's Petroleum Industry, The Case of Bulgaria*. London: GMB Publishing Ltd.

Weithmann, M. W. (1996). *Balkán: 2000 let mezi východem a západem*. Praha: Vyšehrad, spol. s r. o.

CHAPTER 14

Comparison with Russian Operations in the Sector of Natural Gas: The Case of Gazprom

Although closely intertwined for most of their history, the oil and gas sectors differ in several significant respects. Indeed, the gap between them has in recent years become even wider, mostly due to new sources of supply, expanding infrastructure, changing patterns of marketing, and overall growth in the utilization of natural gas. Though the role Russia plays differs in each of these sectors, it does enjoy a position of prominence in both, and over the past 15 years energy-related disputes have arisen between it and several European countries, a number focused on natural gas. They include the 2006 and 2009 gas crises, disputes over gas supplies to Bosnia and Herzegovina, Moldova, or Serbia, and rows over reverse flows to Ukraine in 2014, to name but a few. This, together with the peculiar structural nature of the sector which—especially in the region under discussion—still rests on rigid relationships, may promote the

---

This chapter partially builds on and develops books entitled *Politicization in the Natural Gas Sector in South-Eastern Europe: Thing of the Past or Vivid Present?* (Jirušek, 2017), *Challenges and Opportunities of Natural Gas Market Integration in the Danube Region. The South-West and South-East of the Region as Focal Points for Future Development* (Jirušek & Vlček, 2017), and *Energy Security in Central and Eastern Europe and the Operations of Russian State-Owned Energy Enterprises* (Jirušek et al., 2015), all published by Masaryk University Press, and a paper entitled 'Russia's Energy Relations in Southeastern Europe: An Analysis of Motives in Bulgaria and Greece' (Jirušek, Vlček, & Henderson, 2017), published in the Taylor & Francis journal *Post-Soviet Affairs* (www.tandfonline.com).

© The Author(s) 2019   211
T. Vlček, M. Jirušek, *Russian Oil Enterprises in Europe*,
https://doi.org/10.1007/978-3-030-19839-8_14

politicization and misuse of supply patterns. Unlike in the oil sector, there is just one company involved in exporting natural gas to the region. Gazprom, the state-controlled gas giant, is the exclusive supplier of Russian gas, shipping the commodity through pipelines. As such, it is the dominant foreign supplier in the region. At the same time, Gazprom is the most frequent target of charges that Russia is using state-controlled energy companies as vehicles of foreign influence, leveraging its position in the region along with the structure and features of the natural gas sector in general.

To provide a comprehensive picture of Russian energy-related activity in South-Eastern Europe (SEE), this chapter will offer a brief assessment of Russia's conduct in the natural gas sector in the region under scrutiny.[1] To allow comparable outcomes, the chapter utilizes the same methodology employed in analysing the oil sector and seeks to answer the same research question adapted to the natural gas sector.[2] This involved reformulating the research question as follows: 'Do Russian state-owned energy companies in the natural gas sector in SEE act as tools of the Russian state and serve as vehicles of Russian foreign policy?' For purposes of this portion of the research, the authors have made use of the same theoretical basis and analytical model (i.e. the ideal type model of strategic behaviour), including features and indicators isolated from the data collected. Because the natural gas sector in general and natural gas marketing in particular has evolved in the region in ways that are at variance with what has happened in the oil sector, the authors added an indicator to capture the nature of supply contracts in any given country. Specifically, the concern was whether a take-or-pay condition[3] was present in such contracts. Since this chapter is designed as a brief overview to provide insight into what is undoubtedly a complex environment, the text which follows is limited to key points and findings. The goal has been to put the findings of oil-related research into perspective and sketch a broader picture of the conduct of Russian energy companies in the region.

---

[1] In addition to the states analysed in the rest of the book, the research underlying this chapter included Moldova because of the pronounced role played by the country's natural gas sector and the degree to which it is intertwined with the supply infrastructure that brings Russian gas to the region.

[2] The field research and semi-structured interviews were used also in this research.

[3] That is, the condition securing the supplier a stable flow of funding no matter the real amount of gas consumed by the customer.

14.1   The Sector of Natural Gas: Specific Features Facilitating Misuse

The natural gas sector possesses specific features that make it well-suited to misuse as a pressure tool. These features are embedded in its very physical makeup as well as in the sector's historical development. Natural gas, unlike, oil, cannot simply be loaded on a truck, train, or ocean-going vessel and shipped almost anywhere. Because of its gaseous nature, it must be contained within a sealed space, and this applies to both shipping and storage. This is the fundamental reason why a truly global market for gas has not yet emerged.[4] The most suitable way to ship natural gas is thus to send it via pipeline.[5] However, because it is extremely costly to build a pipeline to transport gas from the areas where it is extracted, suppliers usually take measures to offset their high construction costs. Historically, this has resulted in the establishment of long-term relationships between supplier and customer that provide the supplier with the certainty needed to assure the economics of the project as a whole. It is for these purposes that long-term contracts with take-or-pay conditions (securing a stable flow of funds) were introduced, as were also the prohibition on gas resales and the linking of gas prices to those of oil.[6] These measures naturally gave suppliers the upper hand, allowing them to manipulate prices according to their needs. Moreover, such moves effectively blocked the evolution of a truly functional natural gas market. Instead, the sector has taken on the structure of a series of isolated islands bound to individual supply routes and suppliers.

In any case, the combination of dependency on a single supplier (or even a single supply route) and the lack of other sources[7] of supply, combined with the importance of natural gas for vital subsectors (mainly housing, industry or power generation), make natural gas a perfect tool of

---

[4] Although admittedly LNG has been changing the environment, making overseas commodity exchange a viable option, the market remains generally regionalized, with the majority of gas supplies still shipped through pipelines.

[5] Although LNG has become widely available, since this book is focused on SEE, we will leave it out of the discussion, since it is not an alternative for most countries in the area.

[6] Because of its limited scope and frequent association with oil production, the natural gas price has mostly been derived from the oil price (Jirušek et al., 2015, p. 386).

[7] Imports of liquefied natural gas (LNG) might be an answer to dependency on a single supplier; however, construction of the needed facilities (i.e. liquefaction and regasification facilities) and transport, along with generally missing infrastructure, makes LNG a rather unviable choice for most SEE countries.

leverage, should the pertinent actor decide to employ it as such. The structure of the economies and the state of supply dependency of the states within the region provide a solid incentive to undertake an examination. For most SEE countries, Russia is a major supplier. For five countries within the region (namely, Bosnia and Herzegovina, Bulgaria, Macedonia, Serbia, and Romania), Russia is the sole supplier of imported gas. It is thus no wonder that the region was among those most impacted by the infamous 2009 gas crisis in January of that year. As grave as its impact on the region, it must be noted that not much changed in terms of supply security, and the region is still considered to be among the most supply-vulnerable regions in Europe.

Before we delve into an actual description of Gazprom's behaviour, it is important to familiarize ourselves with two major factors shaping the Russian gas giant's overall strategy—first, the particularities of the region that influence the utilization of natural gas, and then changes to the natural gas sector in countries implementing the rules of the European Union's Internal Energy Market.

14.2   Specifics of Natural Gas Utilization in SEE

Although the utilization of natural gas as an energy source varies among the countries of the SEE region, its importance must be recognized. That importance is determined by factors described below.

14.2.1   Structure of Economies of Countries in the Region

The post-communist states of SEE are characterized by the conspicuous role played by the industrial sector in their overall economic output. This economic structure was considerably influenced by the Soviet model and its use of central planning, which focused on industrial production. Because of this, the economies of the former communist states in SEE (as well as those in other parts of Europe) are highly energy intensive (i.e. a relatively large amount of energy is needed per unit of production) (Ürge-Vorsatz, Miladinova, & Paizs, 2005). As natural gas is often used in the industrial sector, it becomes a crucial factor in industrial production and one on which a supply curtailment might have a severe impact, both on the industry itself and subsequently on the economy as a whole. A similar impact might be felt on households and heating systems, ordinarily another leading consumer of natural gas. Households, as consumers of natural gas, have one additional feature that must be taken into consideration in analysing the potential impact of a supply curtailment. Although they may constitute a relatively small percentage of total energy consumption in a country, that percentage may nevertheless imply that thousands of people will suffer in the cold. As such, this is an unacceptable cost for any government and thus a reason gas supply interruptions remain a significant concern even in countries with relatively low gas consumption. Naturally, viewed from the supplier's perspective, this offers tempting leverage with which to pressure a dependent customer into accepting its conditions.

14.2.2 State of Development of the Natural Gas Sector in SEE

Compared to Western Europe, states in the SEE have far sparser natural gas infrastructure, hindering the development of natural gas and preventing its greater utilization. In most cases these states rely on a single pipeline that brings gas from a sole supplier (Russia). From the energy security perspective, this makes these states extremely vulnerable to supply curtailments. Also, as these countries are among the poorest in Europe, the sector suffers from underfinancing, which further hinders development. It is a fact that in terms of size and consumption, the natural gas sector of the SEE states cannot match that of Western Europe. Apart from formative events on the grand scale like the 2009 gas crisis, the natural gas sector in the SEE countries has mostly remained on the back burner, sidelined in favour of topics like general economic output and crises, ethnic issues, criminality, and recently migration. The overall consumption and relative share of natural gas in these countries' energy mixes are relatively low. Apart from Romania, consumption levels in other states in the region sit at around 2–3 bcm per year (Jirušek, 2017, pp. 56–64). Low relative utilization of natural gas is especially common with the countries of former Yugoslavia which, apart from Slovenia and Croatia, also suffer from poor diversification. Apparently, increasing utilization is dependent upon the availability of gas and the demand in pertinent markets. However, it appears that construction of the local and transit infrastructure needed for higher utilization faces several obstacles. In fact, the problem resembles a vicious circle which is displayed in Fig. 14.1.

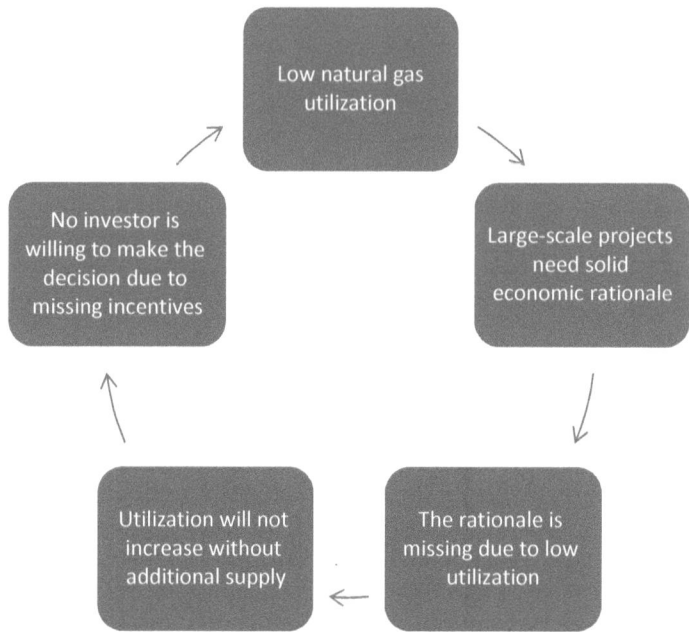

Fig. 14.1   Obstacles to gas utilization in SEE. Source: Martin Jirušek

Demand in the region is inadequate because too few customers are connected to the grid in the individual countries, lowering consumption. Consumption will not increase until more consumers can get affordable gas, and this demands that domestic infrastructure be built and that more gas be brought into the country via expanded or additional transit infrastructure. However, the building of new infrastructure requires financing that is challenging to find, because the low utilization levels do not justify the investment. Plainly said, the region lacks incentives for infrastructural projects that would change the situation, both internal and external. Anchor loads (the threshold for making projects viable) for supply lines that would change the situation are currently higher than demand, as are the anchor loads for new LNG terminals (Jirušek, 2017, pp. 56–64). Needless to say, without this infrastructure, the SEE will remain a region of segregated markets, small in size and unable to shake off the dependency on a single supplier.

14.3 Changes to the Environment Due to EU Internal Energy Market Rules

With the introduction of the so-called Third Liberalization Package into the natural gas sector in 2009, the European Commission tightened the rules for the Internal Energy Market, including those that affect natural gas trading (EUR-Lex, 2009). The Package was not directed specifically against Gazprom but targeted more broadly market incumbents that had dominated the market in the past; nevertheless, Gazprom felt particularly in jeopardy.[8] For years, Gazprom relied on the specific principles present in the gas sector. These principles enabled Gazprom to cement its control over the market and potentially also to provide the company with power to exert pressure on its customers. These principles included the above-noted long-term contracts, destination clauses and a prohibition on reselling gas, the linkage (indexation) noted of gas prices to oil prices and control over transit infrastructure (Jirušek et al., 2015, pp. 384–388). With these measures, Gazprom was able to secure a long-term relationship between itself as supplier and its customers by keeping the market partitioned.

In introducing measures aimed at improving market flexibility and liquidity, the European Commission challenged those principles that had helped Gazprom to maintain its position. The changes that undermine Gazprom's position the most are the ownership unbundling principle, the third-party access principle, and the prohibition of destination clauses. The first principle prohibits any entity from acting as producer and/or supplier and infrastructure owner at the same time. This principle is based on the assumption that such a setup may prevent fair competition and eventually harm the consumer. Therefore, division according to certain legally permitted schemes is required. The second principle requires that equal entry to the market should be available to anyone who wishes to enter, and no one should be prohibited from doing so by being exempted from using certain infrastructure. Lastly, the third principle rests on the assumption that no one should be told what they cannot do with gas purchased, for example, reselling it to other entities (i.e. states). In essence, these rules were imposed to ensure fair competition in the natural gas market (Jirušek et al., 2015, pp. 382–388; Talus, 2011). As for the long-

---

[8] The frustration could be felt as well at the level of Gazprom's main shareholder, the Russian government, as demonstrated by Vladimir Putin at the EU-Russia summit in December 2012 (Euractiv.com, 2012).

term contracts, they are not prohibited as such, but are under increased control by the European Commission (ibid., 201, p. 385; Talus, 2011) as well as increasing pressure from the changing situation on the market where the competition is steadily growing. On a similar note, the European Commission cited oil-indexing as one of the causes of unfair pricing.[9]

These rules are in place within the EU's Internal Energy Market, and any state wishing to join the market must subscribe to them, as well. Here, the Energy Community serves as the main platform within which member states approximate to the Internal Energy Market and gradually implement the main Internal Energy Market (IEM) principles. All countries under examination thus either have already subjected themselves to the rules as EU members, subscribed to these rules as members of the Energy Community, or are in the process of doing so.

It is fair to point out that Gazprom was not alone in utilizing these traditional gas marketing tools. They were also used by other suppliers, basically to offset the cost of infrastructure and to secure a certain level of stability and economic viability in the market.[10] In the same way, as noted above, the EU legislation was not aimed exclusively at Gazprom but rather adopted as a systematic change to harmonize the playing field within the Internal Energy Market. In the region under examination, however, where Gazprom has held a prime position, the predominant impact of the new rules has definitely fallen on the Russian gas giant. Essentially, Gazprom's position has changed in that it is no longer the creator of the environment. Rather, the company is now subject to rules imposed by a superior authority—the European Commission—which is now in charge of the market, at least in countries where the above-mentioned liberalization principles are in place (Jirušek et al., 2015, pp. 384–388).

14.4   Findings

14.4.1   Gazprom Adapts to the Local Environment

Accusations that Gazprom has misused its position to force consumers to follow its bidding appear to be exaggerated in the majority of cases. Generally speaking, Gazprom follows the rules, although it does stretch

[9] The questioning was based on accusations that Gazprom misused its position with the group of Central and Eastern European states (Jirušek et al., 2015, pp. 384–388).

[10] In fact, these measures were used by virtually all major suppliers in Europe, as well as in other markets (e.g. the USA).

them to the maximum extent permissible. From the standpoint of the individual countries, Gazprom does only what it is allowed to do. That means if a country is unilaterally dependent on supplies from Russia, with no alternative supply route, it is highly probable that Gazprom will use its leverage to the fullest. But this need not mean politicization of the relationship. Contrary to what is frequently the popular understanding, in most cases an economic rationale may be found behind Gazprom's conduct. Although using the company's dominance in the marketplace to pressure dependent countries in price negotiations, infrastructure building, and so on, may not be moral, it is not strictly speaking unusual, and it may still be justified on the basis of an effort to make as great a profit as possible.

14.4.2 Gazprom's Approach to Individual Countries Is Determined and Influenced by Several Specific Factors

The research proves that there are several factors that affect and mould Gazprom's (or Russia's, where relevant) behaviour in these countries. These are described below.

14.4.2.1 State of Diversification

Gazprom's monopolistic position in most SEE markets evolved in an era when Russian gas was the only available option. That being the case, it was clearly Gazprom that dictated the contractual conditions. This situation began to change after the fall of the Soviet Union in the 1990s, when some former communist countries reoriented their energy policies and diversified their import portfolios. In those countries that managed to diversify and therefore are no longer dependent upon Russia for 100% of their gas, Gazprom has kept a low profile. They may not be charged the lowest price, but gas deals are usually conducted under a 'business as usual' scenario with little if any dispute over pricing or the stability of supply. Slovenia, Croatia, and Romania are paradigm cases.

Slovenia managed to diversify its supply routes soon after it gained independence and has gradually switched over from traditional long-term contracts to market (hub)-based trading and supplies. With diversified supply routes and firm adherence to EU legislation, there is little room for nonstandard conduct. Croatia presents a similar example. Additionally, the country is poised to increase its own natural gas production. Romania is unique in that its mutual relations with both the Soviet

Union and successor Russia have been chilly, and this has affected many areas of the country's economy, the energy sector included. Over the last decade, the country has managed to substantially decrease its dependence on foreign gas supplies and has been simultaneously developing its domestic resources. It also does not buy gas directly from Gazprom but through intermediaries, which to a large extent prevents any personal involvement by Russian officials or direct linkage of the Romanian foreign policy discourse to gas prices, and so on. On the other hand, this has introduced a great deal of nontransparency into the mix.

14.4.2.2   EU Membership and the Implementation of IEM Rules

Chiefly by means of the Third Liberalization Package, the EU has imposed a set of rules within the Internal Energy Market that have effectively undermined the traditional marketing strategy of the market incumbents that dominated the gas market in the past. Although not aimed squarely at Gazprom but rather more generally at decreasing market fragmentation, increasing competition, market flexibility, and liquidity in the traditionally rigid gas market, Gazprom has been one of the most impacted actors partly but not exclusively because of its position in the Eastern European markets. By its actions, the European Commission effectively challenged the key principles that were the stable pillars of Gazprom's strategy for years and had helped to cement its position in the target markets. Those key principles included long-term contracts, destination clauses (territorial restrictions prohibiting the reselling of gas and, in effect, partitioning the market), linkage of gas prices to oil prices, and control over transit infrastructure. All of these measures were either prohibited or limited by EU legislation.

In countries that implemented the legislation, a substantial amount of strategically motivated or nontransparent conduct was ruled out, thereby restricting the manoeuvring room available to politicize the supply. It is therefore safe to say these rules serve as a 'buffer' between Gazprom and its customers, defining strict rules under which contracts can be conducted. Although the opportunity for Gazprom to implement its traditional marketing tools was significantly diminished by the principles of the Third Energy Package, there is no guarantee that such conduct will not arise; the evidence from other European markets also shows that when the rules are enforced, they form a significant barrier against the Russian company's typical marketing strategy.

14.4.2.3 Foreign Policy Discourse and Political Culture

The foreign policy discourse and political culture of a particular country also force Gazprom to accommodate its strategy to local conditions. While the particular tools and measures used may be broadly similar, experience suggests that Gazprom is capable of adapting its strategy and of applying a case-based approach built around the foreign policy discourse of the country, the existence of areas of potential mutual cooperation, common positions on some issues, and the existence of relevant opinion groups or individuals inclined to cooperate with Russia on a more individualized level. Therefore, if a country's foreign policy discourse towards Russia is lukewarm and firmly anchored in multilateral commercial regimes, as is the case with Romania, Slovenia, and Croatia, politicization, conditionality, or individualized contracts in gas deals are less likely to be a factor. Unfortunately, depoliticized supply contracts constitute an unrealistic goal for most SEE countries. The countries noted are practically the only ones diversified enough to keep their relationship with Russia free of politicization. In the remaining countries, mostly to a great extent dependent on Russian supplies, twinning of energy supplies and political bargaining has been evident.

Still, however, there are exceptions among countries that are EU members, subscribe to the Internal Energy Market rules, and have a diversified import portfolio but are nevertheless prone to politicizing energy-related deals. Most significant is probably Greece, where Prime Minister Alexis Tsipras did not hesitate to use the opportunity to discuss energy-related deals, including gas prices, with Vladimir Putin, no matter the West-imposed sanctions. In this case, however, the specific conditions afflicting Greece's economy, the country's cultural links to Russia, and the personal ambitions of Alexis Tsipras must be taken into account. Bulgaria is another interesting case, ambiguous in that it is inclined towards individually tailored deals, but its closeness to Russia is closely tied to the particular political party in power. Under leftist governments, the stance towards Russia has generally been friendlier and various deals have been signed; right-wing governments have ushered in an era of colder relations in which several disputes have arisen. The best example in this regard is the South Stream project, which was agreed upon during Prime Minister Sergei Stanishev's socialist government and was then disputed and eventually abandoned under the right-wing government of Prime Minister Boyko Borisov.[11] Ultimately, Russia and Vladimir Putin accused Bulgaria of being one of the culprits behind the project's failure.

---

[11] Although Borisov cannot be labelled an outright anti-Russian, which can be seen from his generally pragmatic attitude and several recent steps aimed at improving relations with Russia, he is basically a pro-European politician. This stance obviously placed some boundaries around his foreign policy stances.

*14.4.2.4   The Existence of Intra-State Disputes and Issues that
Translate into Foreign and Energy Policies*
The research concluded that the existence of intra-state disputes, territorial divisions, or cleavages within the society is often translated into energy-related deals. Such issues usually undermine the state's ability to conduct functional energy policies and needed sectoral reforms in the first place. They also often place the state in an unfavourable position on related deals.

In Bosnia and Herzegovina, Russia's unbalanced attitude towards the entities that make up the state further broadens the division between them. As was demonstrated on several occasions, Russia supports the Serb-dominated *Republika Srpska*[12] and has had much more active relations with this entity (and not just to do with energy policy), thereby underlining the unbalanced situation within the federation.

In Moldova, Gazprom tolerates the poor payment morale of customers in the disputed Transnistrian region, where the country's main power generation capacity is located, and charges payments to the government in Chisinau. Such behaviour not only complicates the country's economic situation as one of the poorest economies in Europe, it also helps ossify the complicated situation regarding the separatist region of Transnistria.

14.4.2.5   Predisposition to Agree on Individualized Deals
The reasons certain states are more likely to close individualized or outright nonstandard deals (with both positive and negative implications) vary and may be rooted in:

- unique/nonstandard relations with Russia based on history or cultural ties
- external conditions
- specific personal links and individual aspirations

The first of these options is evident with Serbia, which is perceived as Russia's traditional protégé in the Balkans. This provides a firm basis for Gazprom's actions in the country. But despite this basis, it has not translated into more favourable conditions for gas deals.

---

[12] This is related to Russia's general self-portrayal as the protector of Serbs. Although not really grounded in reality, this is also a basis for the current good state of relations with Serbia itself.

The second option is a factor in the case of Bosnia and Herzegovina, where Gazprom has been unwilling to seal a long-term contract due to the debt the country accumulated during the war in the 1990s and the poor payment morale that ensued. A similar issue is present with Moldova, which has been forced to sign repeated extensions of a long-expired contract because of Gazprom's unwillingness to offer a new one. The sticking point is Moldova's desire to implement the Internal Energy Market rules, thereby weakening Gazprom's market position.

Sometimes, these first two options may be combined. This is what happened with Greece, where good relations and cultural closeness facilitated deals when Greece was forced to deal with a serious economic crisis and Russia's relations with the West had deteriorated.

The third factor was also in evidence to some extent in Greece when Prime Minister Alexis Tsipras held talks with his Russian counterparts to improve the country's bargaining position vis-à-vis the EU during the financial crisis. And it was very much apparent in the case of Bulgaria, where important deals and related legislation were usually adopted during the reign of leftist governments, and where corruption accusations also arose when the same governments were in power, tied to politicians in relation to the Russian-sponsored South Stream project. In Bosnia and Herzegovina, the personal affinities of some representatives of *Republika Srpska* and Russian support for ethnic Serbs in the region facilitated mutual relations between the entity and Russia, clearing the way for several specific deals in the energy sector. Finally, in Moldova, the election of pro-Russian President Igor Dodon facilitated prolonging the previous interim supply deal for an additional three years (Jirušek & Kuchyňková, 2018).

14.4.2.6 Debt Owed to the Supplier

It appears that unresolved debts are very likely to become a point of pressure, one Gazprom usually does not hesitate to employ. Since the majority of states in the SEE are rather poor and their transition periods have been complicated, unpaid debts are not unusual. As may be expected, the issue of unpaid debt is usually brought into the discussion during negotiations over new contracts and the prices paid for deliveries. In line with the point 'Russia favours revenues over good relations' (see below), it is no surprise that threats of supply cut-offs are used also during negotiations with states, like Serbia, that generally have good relations with Russia.

14.4.3   Gazprom Is Motivated to Maintain Its Position Even in a Changing Environment

From the broader perspective, Gazprom's position in Europe has been challenged on several levels. In the Eastern region, its former role as sole supplier has been eroding as a number of states in the region diversify their import portfolios and acquire natural gas from non-Russian sources, as well. The traditional Gazprom trade strategy based on long-term take-or-pay contracts, oil-indexation, and destination clauses has effectively been challenged by EU legislation imposed within the borders of the Internal Energy Market. Therefore, countries implementing these rules find themselves in a more solid position, one that gives them leverage in gas-related negotiations. In transportation, LNG is becoming more relevant, enabling a rising number of countries to diversify their import portfolios. At the same time, LNG is increasing competition in the market and hub-based trading puts pressure on gas pricing and the traditional long-term contracts. In addition, the infrastructure used in the Eastern Bloc, which is much sparser than in the West, is nevertheless getting denser, allowing countries more diverse supply options. Although development in the SEE might be considerably slower than in the more Westerly parts of Europe, based on the number of infrastructural plans that have emerged in the recent decade, it is safe to assume that the trend towards growing interconnectivity is firmly established in this region, as well.

In light of this situation, Gazprom has confronted substantial changes to its formerly stable position. Although the company still plays a major role as a prominent European gas supplier, it is safe to assume that, with these changes, it will focus on retaining as much of its formerly dominant position as possible, and the SEE is no exception. Since the SEE region comprises countries with mostly undiversified gas import portfolios, less developed economies, and somewhat less stable foreign policy discourses, it is likely that this region will be an indispensable part of the company's effort to retain its position. The evidence of Gazprom's behaviour on other European markets so far only confirms such assumptions. It should be noted, though, that Gazprom is no longer the creator of the market but, now that it has been forced by the European Commission to follow Internal Energy Market rules on pertinent markets, has become subject to it.

In terms of Gazprom, the goal of maintaining market position helps explain certain strategic decisions the company has undertaken in past years, as well as its tactics in individual states. This particularly applies to the company's plans to expand its export infrastructure. The South Stream pipeline is a stunning example of this. The project was questionable from an economic point of view in terms of both profitability and the very need for the additional capacity it would have brought. Over the short- to mid-term horizon, the project would probably have made little sense. But from a long-term perspective, it might have provided a firm basis for expansion in the Balkans and Central Europe, where more promising markets are at hand. The strategy also makes sense with regard to Gazprom's reluctance to use the transit infrastructure through Ukraine. New infrastructural projects would thus enable important customers to be reached without complicating disputes. It is also no wonder that the majority of actions of this type have taken place over the last decade when one considers the development of the EU energy markets and related legislation that effectively undermines Gazprom's once sovereign position. From this perspective, Gazprom's activities may be perceived as politically motivated, granting that the policy is to maintain its position on European markets. Therefore, as much as it may have seemed economically unsound in the short to mid-term, from a long-term perspective, the idea of building additional pipelines to reach important markets may eventually turn out to be beneficial. Having this in mind, it is no wonder that Gazprom, with Russia as its biggest shareholder, put such effort into the South Stream project. The frustrating outcome of the endeavour, which was cancelled in late 2014, did not prevent Gazprom from making an attempt to use the Southern route to supply Europe. It ultimately came up with a more modest solution in the form of the Turkish Stream pipeline. Nevertheless, at the time of writing of this text (September 2018), the Turkish Stream pipeline was progressing, with a fair chance of ultimately replacing a good deal of the natural gas currently shipped southwards through Ukraine. Similar notions to do with cementing a position are discernible behind the Nord Stream I and II.

14.4.4   In Most Cases, Gazprom's Behaviour May Be Economically Justified

Contrary to popular perception, there is an economic rationale behind most of Gazprom's actions. Even Serbia and Moldova, which may appear to be clear instances in which political pressure was exerted, may be actually explained on the basis of ordinary supplier-consumer relations. The threats, ultimately realized, of a gas supply cut-off to Serbia in 2014 may be attributed to the debt Serbia owes Russia rather than any effort to steer the country in a particular direction. In the case of Moldova, resistance against the country's efforts to implement the EU's internal market rules is also understandable given Gazprom's stance towards EU rules in general. This is in any event not to claim that these activities are acceptable or even absolutely legal, but they do at least make economic sense. Additionally, cases of similar behaviour by Gazprom may be found in other Central and Eastern European countries.

14.4.4.1   It Is Not the Measures That Are Suspicious, But Rather the Timing and the Context

Despite what was said in the previous paragraph, Gazprom's behaviour has not always been crystal clear. Although the steps it has taken may have an economic rationale behind them, when it comes to timing, this is not always the case. With Greece, for example, a substantial natural gas agenda was brought to the table when it was advantageous for Russia to get even closer to Greece in 2015. In Bulgaria, a gas price discount was offered in exchange for signing a bilateral deal and accelerating preparations for the South Stream project in 2012. In Serbia and Bosnia and Herzegovina, the threat, ultimately realized, to cut off the gas over the gas debt in the second half or last quarter of the year clearly put pressure on debtors as winter loomed. In the case of Macedonia, Russian Foreign Minister Sergey Lavrov did not lose a minute blaming 'external forces' for orchestrating public unrest in the country as punishment for Macedonia's consent to take part in extending the Turkish Stream. And in the case of Moldova, a steep increase in the price of gas correlates with a rapid worsening of mutual relations after 2003. The series of economic sanctions imposed on Moldova also correlates with the signing of the Association Agreement between Moldova and the EU.

14.4.5   Gazprom Is Still a 'Normal' Company

Gazprom is first of all a profit-seeking corporation. Also, Gazprom and the Russian state, its majority shareholder, are two separate entities. Their interests may be largely similar but on some occasions they do markedly diverge. Gazprom itself might in these instances thus face pressure from the Russian state to behave in a manner that contradicts an economically sensible, profit-oriented logic. This pressure may be exerted for foreign policy reasons or for reasons to do with internal affairs.

14.4.5.1 But at the Same Time, It Is a State-owned Enterprise and Easy-to-Use Tool

The fact that the Russian state is a majority shareholder of Gazprom and that the company is the crucial supplier of gas to countries that were historically and often politically bound to Russia cannot be neglected. Also, as Russia is vitally dependent on exports of natural commodities, any gas supply deals are very closely monitored by the government. The importance of gas exports to the Russian economy further explains potential state interventions and support for both Gazprom and the supply deals it administers. From the consumer perspective though, it ultimately makes no difference what motivation lies behind the pressure exerted, be it political or economic.

Nor is it a one-way relationship. Given their historical roots, certain social and political groups in the SEE countries are still prone to cutting deals with Russia regardless of the potential impact on their own energy security or Russian foreign policy discourse. It is thus little wonder that shady deals and instances of bribery occur in such an environment.

14.4.6 Russia Favours Revenues over Good Relations

Despite the popular perception that Russia, through Gazprom, rewards its allies with favourable prices to cement their mutual relations, the research showed the reality is otherwise. Even states whose relations with Russia are good or who are perceived to be Russian protégés are not safe from the threat of supply cut-offs or pressure to pay a high price. Serbia, Greece, and Macedonia are good examples of the price issue,[13] and Serbia and Bosnia and Herzegovina, too, have twice been threatened with cut-offs over the past decade, despite Russia's long-term support for all Serbs. Moldova had a similar experience, although in this case, threats aimed at weakening the country's will to join the Internal Energy Market were tacit instead of official.[14] Although the price is often influenced by the rather nontransparent structure of intermediaries frequently present in the

---

[13] Serbia also experienced supply cuts attributed to its debt in 2014.
[14] Moldova's will to join the Internal Energy Market was problematized mainly by then Russian Deputy Prime Minister Dmitry Rogozin.

228 T. VLČEK AND M. JIRUŠEK

Balkans, Russia definitively does not use Gazprom to offer these countries advantageous conditions as a reward for their support. When it has appeared to, it has usually been part of a barter-like deal, as was the case with Greece. There, at a time when Vladimir Putin needed European allies to erode EU unity on anti-Russian sanctions, negotiations were opened on gas-related issues that touched on a construction loan for a pipeline over Greek soil. An analogous situation involving the offer of favourable conditions in exchange for concessions arose in Bulgaria in 2012 (see above).

It would thus seem that the aim is to maximize revenue above all, and this is understandable given the rather low overall utilization of natural gas in the SEE states. It is also worth noting that the Balkan states were the most severely hit during the 2009 gas crisis. Russia did not hesitate to sacrifice supplies to the SEE over disputes concerning unpaid Ukrainian bills.

As hinted at above, in looking at the price map of Gazprom's European customers there seems to be no tie, contrary to popular perception, between a good relationship with Russia and low prices. On the contrary, there are several instances in which states who have fairly good relations with Russia pay very high prices for gas deliveries. Serbia, Macedonia, and Greece are good examples of this. Although it is probable that the delivery price reflects the current state of relations in some cases to some extent, the effect is indirect. The degree of dependency, the number of alternative suppliers, and the general transparency of the sector all come into play. Therefore, claiming a clear correlation between a particular country's relationship with Russia and its gas price would likely be an oversimplification.

14.4.7   Intermediaries as an Important Factor Increasing Nontransparency

Generally speaking, states in the region suffer from nontransparency and corruption, and the natural gas sector is no exception. Quite often a nontransparent network of intermediaries and middlemen is part of the supply chain, complicating it and bringing it under suspicion. These intermediaries usually operate supply contracts, generally in countries where the price of gas is not disclosed. The ownership structure of these intermediaries is generally nontransparent and, as the field research demonstrated, their presence is implicated as one reason for the high price of gas paid by these countries. The evidence shows these intermediaries usually have ties to

Gazprom and its subsidiaries, or to individuals linked to Gazprom or the Russian bureaucracy. A clear example is Serbia, where the company Yugorosgaz operates as an intermediary, reselling Russian gas to state-owned Srbijagas while simultaneously operating the infrastructure in the Southern part of the country. Yugorosgaz is a joint venture of Gazprom, Srbijagas, and the Centrex Group, the latter related to Gazprom's operations in several European countries including the Czech Republic, Slovakia, Hungary, Italy, Switzerland, Austria, and Great Britain. Other examples of nontransparent ties leading to Russia are Macedonia and, to some extent, Romania.

14.4.8 Involvement by Russian Officials Is an Important Indicator

As indicated above, Gazprom would like to be seen as a reliable business-oriented company, but the evidence shows that involvement by Russian state officials is common, and this does not fit into a 'business-as-usual' scenario. The Russian state may be a majority owner in the company, but its involvement may nevertheless raise a red flag in some cases. The research showed that in countries that depend upon Russian sources for all or almost all of their imports, Russian officials often get involved in energy-related deals. There is naturally nothing unusual about politicians backing state-owned companies, especially those that are highly important to the mother country in the way Gazprom is to Russia. In the cases under scrutiny, though, Russian officials acted more like regular Gazprom representatives, negotiating the specific terms of individual deals. This 'substitutability' of Russian for Gazprom officials is symptomatic. In addition, the presence of Russian state officials usually correlates with the importance a particular country has or might have for the transit of Russian gas or Russian foreign policy goals in general. The willingness of a state to agree individualized deals of the type described above naturally plays a key role.

It also demonstrates that the above-mentioned type of deal is usually not struck in countries whose gas supplies are well diversified, or that implement the principles of the Internal Energy Market, or, rather, both. Such states typically have fairly standard, business-oriented relations with Russia and do not build on personal linkages between state officials. Romania, Slovenia, and to some extent Croatia are good examples in this regard. Romania boasts low relative import dependence and has no standard long-term contract signed with Gazprom. Although the setup under which Romania secures its imports through intermediaries has its disadvantages (primarily in terms of price and nontransparency), it seems to have eliminated politicization at the state level. Slovenia, for its part, managed to diversify its import portfolio soon after gaining independence by building an interconnector to Italy. This enabled it to obtain gas of non-Russian origin from various sources. The country has a pro-Western orientation and quickly implemented the main principles of a market economy. In fact, the natural gas sector currently obtains the majority of its supplies via hub-based trading. The situation thus provides very little opportunity for political manoeuvring or misuse. Slovenia's neighbour Croatia utilizes the same supply route as its Western neighbour and between 2010 and 2013 was able to acquire 100% of its supplies from somewhere other than Russia. Also, the country's pro-Western foreign policy discourse and rather chilly relations with its neighbour, Russia's protégé Serbia, have created an environment unreceptive to non-commercial, strategically motivated conduct. The country's long-term (although economically doubtful) effort to build an LNG terminal also confirms the general discourse of not relying on ties to Russia.

14.5   Conclusion

Generally speaking, the fact that Gazprom behaves in a manner that is sometimes far from 'business as usual' can nevertheless not be used to claim the company is simply a tool of its government or the government's foreign policy. Rather, Gazprom, often with significant support for its homeland government, does not hesitate to seize opportunities to strengthen its position within particular markets. This essentially means that the company mostly follows the rules of the environment while stretching them to the maximum extent permissible. The conclusion here might be that the company is aware of the changing environment within which it operates and tries to shore up its position to the fullest, taking advantage of government support when it is possible to do so. Although every company works to secure its own position, in this instance close attention must be paid to timing, context, and case-specific intervening factors and circumstances in which governmental support is exerted. Since the region is composed of a heterogeneous group of countries whose gas sectors differ from one another, recommendations may be derived for keeping the behaviour of Gazprom and its subsidiaries under control. The key principles that should be implemented are adherence to the rules of the Internal Energy Market, transparency, and legality in business dealings in the sector, and, perhaps most importantly, a diversified import portfolio.

As for the research question, no simple 'yes' or 'no' outcome is available. Instead, the situation proves to be highly case- and environment-specific. To offer a binary answer would therefore be to oversimplify a more complex reality. Addressing the question properly requires that several factors thoroughly assessed above be taken into account. It follows that the most accurate answer to the research question ('Do Russian state-owned energy companies in the natural gas sector in SEE act as tools of the Russian state and serve as vehicles of Russian foreign policy?') would be as follows:

Gazprom, as the company in charge of delivering Russian gas to the SEE region, does appear in some cases to serve as a tool for particular governmental policies. These policies, however, need not necessarily meet the goals of Russian foreign policy but may instead be pragmatically formulated to meet the economic objectives of the company (or the Russian state) to earn as much revenue as possible and to secure its position in the market.

Whatever the intentions behind Gazprom's behaviour, the targeted countries should thus proceed carefully in accordance with the advice given in the previous section.

SOURCES

Euractiv.com. (2012, December 21). *Putin Slams Barroso: 'You Know You Are Wrong, You're Guilty'. EurActiv.* Retrieved from http://www.euractiv.com/section/energy/news/putin-slams-barroso-you-know-you-are-wrong-youre-guilty/

EUR-Lex. (2009, August 14). *Directive 2009/73/EC of the European Parliament and of the Council.* Retrieved from http://eur-lex.europa.eu/legal-content/EN/ALL/?uri=CELEX%3A32009L0073

Jirušek, M. (2017). *Politicization in the Natural Gas Sector in South-Eastern Europe: Thing of the Past or Vivid Present?* Brno: Masaryk University.

Jirušek, M., & Kuchyňková, P. (2018). The Conduct of Gazprom in Central and Eastern Europe: A Tool of the Kremlin, or Just an Adaptable Player? *East European Politics and Societies and Cultures, 32*(4), 818–844. https://doi.org/10.1177/0888325417745128

Jirušek, M., & Vlček, T. (2017). *Challenges and Opportunities of Natural Gas Market Integration in the Danube Region. The South-west and South-east of the Region as Focal Points for Future Development.* Brno: Masaryk University. Retrieved from https://munispace.muni.cz/library/catalog/view/945/2957/679-2

Jirušek, M., Vlček, T., & Henderson, J. (2017). Russia's Energy Relations in Southeastern Europe: An Analysis of Motives in Bulgaria and Greece. *Post-Soviet Affairs, 33*(5), 335–355. https://doi.org/10.1080/1060586X.2017.1341256. Retrieved from www.tandfonline.com

Jirušek, M., Vlček, T., Koďousková, H., Robinson, R. W., Leshchenko, A., Černoch, F., et al. (2015). *Energy Security in Central and Eastern Europe and the Operations of Russian State-Owned Energy Enterprises.* Brno: Masaryk University. ISBN 978-80-210-8048-5. Retrieved from https://munispace.muni.cz/library/catalog/book/790

Talus, K. (2011). Long-term Natural Gas Contracts and Antitrust Law in the European Union and the USA. *Journal of World Energy Law and Business, 4*(3), 260–315.

Ürge-Vorsatz, D., Miladinova, G., & Paizs, L. (2005, May 17). Energy in Transition: From the Iron Curtain to the European Union. *Energy Policy, 34,* 2279–2297. Retrieved from http://ac.els-cdn.com/S0301421505000947/1-s2.0-S0301421505000947-main.pdf?_tid=8d97b93e-6004-11e6-b419-00000aacb362&acdnat=1470948515_6e3dcbb690eda7c586b06d038 26df5ae

CHAPTER 15

General Conclusion

It may be stated that energy is an indispensable factor in relations between Russia and the South-Eastern European countries. Although intertwined for most of their history, the oil and gas sectors demonstrate substantial differences in the roles they play in this relationship. These differences stem mainly from the disparate nature of both commodities and their markets. But there are certain features that emerged from the research that are key embodiments of the attitude taken by the Russian state and Russian energy companies to the region. It must also be acknowledged that whether an energy company is state-owned or privately owned; there is always a strong link to the Russian government, because business on the large scale cannot be done in Russia in defiance of the Kremlin, especially in the vitally important energy sector.

It appears that 'Russian' conduct in the region—both that of the state and that of companies—is predominantly project oriented. This means all actions conducted by a Russian company or by the Russian state via its representatives in either of the two sectors examined are subordinated to the project goal. Ultimately, it is the importance of the project in both political and economic terms that matters. In essence, the project always comes first, closely followed by the adaptation of pertinent policies. Therefore, the general state of relations between Russia and the country in question is not the main motivator of the Russian attitude. Rather, these relations serve as facilitating factors where possible. Given that the individual states in the region are of marginal significance given their oil

© The Author(s) 2019
T. Vlček, M. Jirušek, *Russian Oil Enterprises in Europe*,
https://doi.org/10.1007/978-3-030-19839-8_15

and gas consumption, Russia perceives ties to them to be secondary to projects that generally serve some broader purpose politically and economically.

The research showed that Russia is willing and able to use all kinds of tools and every available opportunity to foster projects it deems important, no matter the environment. Naturally, the environment determines which tools can be used. Generally speaking, nontransparent conditions are favourable for a wide variety of measures, including the personalization of politics, the coupling of energy issues with politics, corruption, and so on.

But as it turns out, and contrary to popular perceptions, it is the regional state and the attitude it takes that determine the behaviour of Russia and companies of Russian origin. If that state is forthcoming or proves susceptible to influence for any reason (e.g. highly one-sided dependence, nontransparent policies, corrupt or opportunistic behaviour on the part of politicians, historically good relations), it is highly likely that Russia and its energy companies will use this opportunity to the fullest.

Although geopolitics may be part of Russia's motivation in the area, we cannot clearly differentiate whether the reasoning behind certain behaviours is purely geopolitical or economic. As Russia is vitally dependent on energy commodity exports, a mixture of both is usually present, with varying content depending on the situation or state. Ultimately, though, it is of little import to the state being targeted why the situation was hijacked and the state pressurized. Because the region is generally susceptible to individualized, opaque deals, the general recommendation would be to focus on transparency and legality to rule out any suspicious behaviour by any actor.

INDEX[1]

[1] Note: Page numbers followed by 'n' refer to notes.

© The Author(s) 2019
T. Vlček, M. Jirušek, *Russian Oil Enterprises in Europe*,
https://doi.org/10.1007/978-3-030-19839-8

MIX

Papier | Fördert
gute Waldnutzung

FSC® C083411

Zeitfracht Medien GmbH
Ferdinand-Jühlke-Straße 7
99095 Erfurt, Deutschland
produktsicherheit@kolibri360.de